OXFORD STUDIES IN PHYSICS

GENERAL EDITORS

B. BLEANEY, D. W. SCIAMA, D. H. WILKINSON

THEORETICAL COSMOLOGY

BY

A. K. RAYCHAUDHURI

CLARENDON PRESS · OXFORD

1979

Oxford University Press, Walton Street, Oxford OX2 6DP

OXFORD LONDON GLASGOW
NEW YORK TORONTO MELBOURNE WELLINGTON
KUALA LUMPUR SINGAPORE JAKARTA HONG KONG TOKYO
DELHI BOMBAY CALCUTTA MADRAS KARACHI
NAIROBI DAR ES SALAAM CAPE TOWN

ISBN 0 19 851462 X

Published in the United States by
Oxford University Press, New York

British Library Cataloguing in Publication Data

Raychaudhuri, A. K.
Theoretical cosmology.—(Oxford studies in physics).
1. Cosmology
I. Title II. Series
523.1'01 QB981 79-40847

ISBN 0-19-851462-X

Printed in Great Britain by
Lowe & Brydone Printers Ltd.
Norfolk

PREFACE

Today cosmology is one of the most exciting and at the same time confusing subjects. The book aims to convey to the reader some of the excitement while not hiding the confusions that remain. Thus while evidences of the microwave background radiation and the abundances of deuteron and helium as bringing support for standard big bang cosmology are dealt with in fair detail, the unsolved puzzle of some apparently anomalous red shifts and the tendency towards a negative deceleration parameter also receive attention. On the formal theoretical side, the Hawking-Penrose theorem on singularities as well as numerous other less-well-known theorems of varied importance are presented, mostly without the details of a proof. The two-tensor theory of Isham, Salam, and Strathdee and the conformally invariant theory of Hoyle and Narlikar are mentioned but, consistent with the fact that there is so far only a meagre literature on these theories, they receive rather short coverage. Simple but reasonably detailed reviews of the Brans-Dicke theory and the so-called Einstein-Cartan theory are also presented. The steady-state theory, by itself, does not find any place although some references to it occur here and there. This reflects the present attitude of the average cosmologist (call it the 'big-bang band wagon' if you like).

Like any other branch of theoretical physics, theoretical cosmology is also subject to the constraints set by observational data. Thus there are frequent references to these in addition to a full chapter on the analysis of observational data.

The author hopes that both the advanced graduate student and the physicist who is not a cosmologist will find in the book a readable and informative review of the subject while the professional cosmologist may take it as a handy reference book on various topics in his subject.

In writing the book I have freely used standard treatises on cosmology and related subjects - indeed there is little in the book that is not to be found in the published literature.

In particular the book reproduces the following figures:

(1) Fig. 3.1 on page 30 from 'Relativity, thermodynamics and cosmology' by R.C. Tolman (Clarendon Press, Oxford), 1934.

(2) Fig. 7.1 on page 139 from the paper 'Big bang nucleosynthesis revisited' by R. Wagoner (Astrophys. J. 179, 343, 1973, University of Chicago Press).

(3) Fig. 4.1 on page 59 from the paper 'Steps toward the Hubble constant' by A.R. Sandage and G.B. Tammann (Astrophys. J. 196, 313, 1975, University of Chicago Press).

(4) Fig. 12.1 on page 215 from the paper 'Lemaitre's Universe and Observations' by N. Kardashev (Astrophys. J. 150, L 135, 1967, University of Chicago Press).

(5) Fig. 13.1 on page 221 from the artic 'Primordial Turbulence and Intergalactic Matter' by J. Silk in Cargese Lectures in Physics Vol. 6, edited by E. Schatzman (Gordon and Breach, New York, 1973).

The author expresses his sincere thanks to the publishers i.e. The Clarendon Press, The University of Chicago Press, and Gordon and Breach for permissions to use these materials.

A.K.R.

CONTENTS

Chapter 1.	INTRODUCTION	
	1. The hydrodynamical approximation	1
	2. The cosmological principle	2
	3. The Robertson-Walker metric	7
Chapter 2.	NEWTONIAN GRAVITATION AND COSMOLOGY	
	1. Formulation and critique of Newtonian cosmology	10
	2. A comparison between Newtonian and relativistic cosmology	14
	3. Can collapse be halted by vorticity?	17
Chapter 3.	GENERAL RELATIVITY AND RELATIVISTIC COSMOLOGY	
	1. Basic equations of isotropic relativistic cosmology	21
	2. The cosmological term and the static universe models	23
	3. The energy-momentum tensor and bulk viscosity	26
	4. The singularity of isotropic models	28
	5. The time behaviour of isotropic models	28
	6. Geometry of the three space	34
	7. The age of the universe	36
	8. Paths of particles and light rays and horizons	37
	9. The red shift	40
	10. Relations involving H, z, and the deceleration parameter q.	42
	11. Relation between the look back time and z, and the age of the universe and H, q	43
Chapter 4.	THE ANALYSIS OF OBSERVATIONAL DATA	
	1. The questions to be settled	46
	2. Proper distance, luminosity distance, and distance by apparent size	48
	3. Relation between the luminosity distance and the apparent magnitude	49

4. Relation between the luminosity distance D, the red shift z, the Hubble constant H_0, and the deceleration parameter q_0 56
5. The number counts or the N - m relations 63
6. The angular measurements 65
7. More general analysis of the observational data 68
8. The curvature of space and the search for missing matter 68

Chapter 5. RELATIVISTIC MODELS NOT OBEYING THE COSMOLOGICAL PRINCIPLE
1. The motivation behind study of these models 79
2. Some general formulae and theorems 80
3. Homogeneous universes 85
4. The Gödel universe 92
5. Spherically symmetric nonhomogeneous solutions 95
6. The role of gravitational waves in the dynamics of anisotropic universes 97
7. The Lichnerowicz universes 98

Chapter 6. THE MICROWAVE RADIATION BACKGROUND
1. Thermal nature of the radiation 100
2. The universality of the radiation 103
3. The isotropy of the radiation 105
4. Alternative ideas about the origin of the background radiation 110
5. Interaction of high energy particles with the microwave photons 114
6. The isotropy of the background radiation, particle horizons, and causality 117

Chapter 7. THERMAL HISTORY OF THE UNIVERSE AND NUCLEOSYNTHESIS
1. Elementary particle and the hadronic big bang 123
2. The early universe 125
3. The reheating of the cosmic gas 133
4. Nucleosynthesis and the abundance of light elements 135

Chapter 8. THE SINGULARITY OF COSMOLOGICAL MODELS

1. The existence of cosmological singularity 145
2. The oscillatory approach to singularity in some anisotropic models 151

Chapter 9. THE GRAVITATIONAL CONSTANT AS A FIELD VARIABLE

1. The large nondimensional numbers and the Dirac hypothesis 157
2. A varying G 158
3. The scalar tensor theory of Brans and Dicke 163
4. Cosmological solutions of the Brans-Dicke theory 168
5. Comparison with observations 174
6. Anisotropic solutions in Brans-Dicke cosmology 175
7. Singularity in scalar tensor theories 180

Chapter 10. COSMOLOGICAL MODELS BASED ON EINSTEIN-CARTAN THEORY

1. The field equations of the Einstein-Cartan theory 182
2. Cosmological models of the Einstein-Cartan theory 185
3. Maxwell equations and cosmological models with a magnetic field 191

Chapter 11. THE COSMOLOGICAL SINGULARITY IN TWO RECENT THEORIES

1. The two tensor theory of gravitation 199
2. The conformally invariant theory of Hoyle and Narlikar 200

Chapter 12. FATE OF PERTURBATIONS OF ISOTROPIC UNIVERSES

1. The basic problem 203
2. The case of Newtonian cosmology 205
3. Perturbations of the Friedmann universe 208
4. Density perturbations in anisotropic models 217
5. Density perturbation in Brans-Dicke cosmology 217

Chapter 13. THE FORMATION OF GALAXIES

1. The primeval turbulence theory 218
2. Other ideas about the origin of galaxies 223
3. An explanation of galactic magnetic fields 224

Chapter 14. BARYON SYMMETRIC COSMOLOGY

1. The Klein-Alfvén theory 227
2. The Ómnes theory 230
3. The observational situation 232

Chapter 15. SOME ASSORTED TOPICS

1. The extragalactic radio sources 234
2. Mach principle 240
3. Olbers' paradox 245
4. Concluding remarks 247

APPENDIX 249

REFERENCES 252

AUTHOR INDEX 281

SUBJECT INDEX 293

1
INTRODUCTION

1.1. THE HYDRODYNAMICAL APPROXIMATION

Cosmology is the science of the Cosmos - of order that exists in the universe at large. At first sight the universe consists of stars, star clusters, and galaxies or the nebulae. The galaxies again form clusters and even superclusters. Besides these, there exist quasi-stellar objects, electromagnetic radiation ranging all the way from the microwave region to hard γ-rays and possibly also gravitational radiation of appreciable intensity, black holes, and matter in other forms which elude direct observation. Indeed it has often been supposed that ionized hydrogen makes up a total mass many times that of all the galaxies together and yet this remains undetectable in radio or optical observations. If this is the case the galaxies are just pecularities in an otherwise continuous distribution and can be disregarded in the first instance so far as their gravitational influence is concerned.

Whatever the actual form and distribution of the constituents of the universe may be, the basic problem of cosmology is the dynamics of the system. In this study of dynamics, the interaction which is of dominating importance is gravitation - the other long-range interaction, the electromagnetic forces, may be disregarded, for matter on a large scale is more or less electrically neutral* and the magnetic fields that are known to be present in galaxies lead to interactions of a much smaller order of magnitude than the gravitational.

The dynamical problem in its perfect generality is

*Bondi and Lyttleton (1959) proposed that an intrinsic difference in the charges of the electron and proton may lead to an electrically charged universe. See however, Hoyle (1960), King (1960), and Barry (1974). Bally and Harrison (1978) have considered the possibility that the universe may be electrically polarized due to the escape of electrons from stars and galaxies - even then the effect of such polarization has been found to be negligible.

clearly insoluble and one has to introduce additional assumptions and approximations. The first approximation that is usually made is that of continuous matter distribution.* If intergalactic hydrogen is the predominant constituent of the universe, then this only means that we are disregarding, for the moment, the galactic concentrations. On the other hand, if the galaxies are the dominant contributors to the total mass of the universe then the approximation is similar to that in fluid mechanics where also one disregards the discrete molecular structure and considers instead a continuous density distribution. The approximation would then necessitate that we deal with regions of large size compared to average galactic separations (10^6 light years).

1.2. THE COSMOLOGICAL PRINCIPLE

The most powerful assumption in standard cosmological theory is that of homogeneity and isotropy, often referred to as the cosmological principle. The ordinary meaning of the assumption is that on a sufficiently large scale the universe has no preferred position or direction, i.e. the picture of the universe as seen from different locations would be essentially the same and also there would be no observable difference between different directions. Obviously, in view of the manifest irregularities one should give a precise definition of what is meant by a sufficiently large scale but this is not easy. In any case, in order to be meaningful this 'large scale' must be small compared to the universe as a whole or that portion of it which is theoretically observable at a particular epoch.

Several different approaches to the cosmological principle seem clearly discernible and we shall name them as philosophical, mathematical, deductive, and empirical. The first group had an early protagonist in Milne (1935) in the thirties and Bondi, Gold, and Hoyle in the more recent past. They regarded the principle as the corner stone of all

* For a statistical approach as alternative to the hydrodynamical approximation, see Ehlers (1971), Sachs and Ehlers (1971), and Trümper (1971).

cosmological theories; the main justification seemed to be an emotional appeal and the fact that, without it, the applicability of the laws of physics, as we have come to know from our observations here and now, becomes questionable in situations where the physical conditions are drastically different. In this way Bondi and Gold (1948) were led to the idea that, not only is there spatial homogeneity and isotropy, but that there is also a uniformity in time, and the universe must therefore be in a steady state - a static state was ruled out by difficulties connected with Olbers' paradox and the observed red-shift of spectral lines from distant galaxies. A field-theoretic background for the steady-state models involving creation of matter, was also proposed by Hoyle (1948) and later extensively studied by Hoyle and Narlikar (1963, 1964). At present, observations on the microwave background, the relative abundance of Helium, the presence of appreciable amounts of deuteron in different locales as well as the radioastronomical data are generally believed to be irreconcilable with the steady-state cosmology so that along with the steady-state theory the philosophical approach to the cosmological principle has lost much of its appeal.

The second group, whom we have called mathematical, regard the cosmological principle as essentially one of great mathematical beauty as with the help of group-theoretic ideas applied to Riemannian geometry, the principle leads to a simple, elegant, and to an extent a unique foundation of cosmology. To them, any observational support in favour of this principle would no doubt be significant as showing that the principle is applicable to the actual universe, but even in the absence of such support they would regard the study of models obeying the cosmological principle a worthwhile endeavour. Indeed one may even subscribe to the idea as suggested by Dicke (1961) that there is not one, but a whole ensemble of all universes allowed by theory - in that case the universes conforming to the cosmological principle would surely be of great interest owing to their beauty and simplicity.

Thirdly, the deductive school would like to deduce the

cosmological principle rather than to assume it - i.e. they would like to demonstrate that starting with arbitrary initial conditions, physical processes lead to an evolution of a universe obeying the cosmological principle - inhomogeneities and anisotropies being smoothed out in a time small compared to the age of the universe. The motivation behind this approach is a logical difficulty inherent in the cosmological principle which was first emphasized by Misner (1969b). The models of general relativity which conform to the cosmological principle have a particle horizon, i.e. at any particular epoch signals can reach a particular point only from a limited region of the universe, so that regions beyond a certain distance have no possibility to communicate if we consider any particular instant. However the hypothesis of the cosmological principle as well as the observed isotropy of the background microwave radiation require a correlation in the conditions at different regions (throughout the universe according to the cosmological principle and over the last scattering surface from the isotropy of the radiation) which are beyond their respective horizons. How are we to explain this correlation?

Misner advocated an anisotropic model at the early stages of the universe and it was claimed that there is effectively no horizon so that a communication between different regions may bring about uniformity. Thus one had the rather paradoxical situation that the cosmological principle holds at present only because it was not true initially in the so called mix-master universe of Misner.

Assuming the early universe to be anisotropic but nevertheless homogeneous in which vorticity was already absent, Misner (1968) apparently showed that neutrino viscosity could reduce the shear from very large values in the early universe to the present low level. However Misner's claim that neutrino viscosity can effectively decrease the anisotropy has been questioned by later investigators, e.g. Stewart (1969), Collins and Stewart (1971), Doroshkevich, Zel'dovich, and Novikov (1968), and Doroshkevich, Lukash, and Novikov (1971, 1973). Again considering homogeneous universes admitting groups of motions of Bianchi types V and

IX (the former has open and the latter closed space-sections) Hawking (1969) arrived at the conclusion that viscosity reduces vorticity, but starting from arbitrary initial conditions viscosity cannot reduce vorticity to the low values deduced from the present observations on isotropy of the background radiation unless the present density of matter in the universe is less than $2{\cdot}5 \times 10^{-33}$ g cm^{-3}. However such a low value of density is clearly ruled out by observations.

More recently Chitre (1973) and MacCallum (1971) have shown that the removal of horizons in the so called mixmaster universe is also open to question. At present, the Misner programme at least in its original form has been practically given up (Misner 1973).

The fourth group has an essentially open mind regarding the cosmological principle and would go by the empirical evidence. However, owing to a variety of complicating factors, the observations as they stood before the discovery of the microwave-radiation background were by no means decisive on this point. A naive empirical criterion to decide whether the principle holds good is to study the distribution of galaxies. Apparently there is a good degree of isotropy in the distribution, although one can observe very few galaxies near our galactic plane and there seems to be a greater number in the northern hemisphere. However, the discrepancies may be explained away as due to the obscuring gas clouds in the plane of our galaxy and the presence of a specially rich cluster in the northern hemisphere. For the distribution in depth, if one assumes Euclidean geometry and stationary galaxies, homogeneity requires

$$\log N = 0{\cdot}6\, m + \text{const.}$$

where N is the number of galaxies brighter than apparent magnitude m and we have assumed that the absolute magnitudes (i.e. the intrinsic luminosities) of all the galaxies are the same. An observational confirmation of the above relation would appear, at first sight, to be a verification of the homogeneity postulate. But the effects of possible departure of the spatial geometry from Euclidean and the influence of Doppler shift on the apparent magnitudes have been left out and the second effect at least is known to be of extreme importance as the distance of the observed galaxy increases. Again a homogeneity of the galactic distribution leading to a constancy of the smoothed out density of matter would not necessarily mean complete homogeneity and isotropy as is conceived in the cosmological principle - the red shift for example may still be anisotropic. Indeed there are well known solutions of the equations of general rela--ivity in which, although the matter distribution is uniform, there is anisotropy of expansion and the geometry at different points is not identical (Bondi 1969; Raychaudhuri 1955b; Thomson and Whitrow 1967, 1968).

There is a school of opinion which holds that there is a definite systematic non-uniformity in the distribution of galaxies. Thus de Vaucauleurs (1970, 1971) concluded that the density distribution can be represented by the formula $\rho \propto r^{-\theta}$ where $\theta = 1{\cdot}17$ and r is the distance from our locale. However an analysis by Sandage, Tammann, and Hardy (1972) contradicts this finding and suggests $\theta = -0{\cdot}115 \pm 0{\cdot}030$, so that any departure from uniformity, even if present, can only be marginal (see also Rowan-Robinson 1972). The de Vaucouleurs law of density has inspired the study of hierarchical models in which a compromise is made between clustering and an over all homogeneity (Wertz 1971; Haggerty and Wertz 1972). The hierarchical idea has apparently been supported by the investigations of Peebles and collaborators (Peebles and Wu 1970; Peebles and Hauser 1974; Peebles 1974a,b). Turner and Gott (1975) have found that for galaxies brighter than 14th magnitude there are two populations - one strongly clustered containing 60 per cent of all galaxies and the remaining ones distributed almost uniformly.

The greatest empirical support in favour of the cosmological principle comes from the observation of the microwave-radiation background. The radiation is isotropic at our locale to an accuracy of ~0·1 per cent (an accuracy far higher than that attainable in other cosmological observations) and this indicates an isotropy of the universe obtaining as far as the last scattering surface which is at least as far as $z \approx 8{\cdot}6$ and may be as far as $z \approx 1000$. (z indicates the red shift $\delta\lambda/\lambda$; as z depends on the distance between the points of emission and observation, it can be used as a measure of both distance and time. The two values of z for the last scattering surface correspond to high and low density universes respectively.) It is difficult to suppose that this isotropy is a peculiarity of our own neighbourhood - in a way that would be going back to the pre-Copernican idea that we are at the preferred position of the centre of the universe. In the spirit of our present age it seems natural to assume that this isotropy obtains at all space-time points. It may then be shown that the equations of general relativity lead to the cosmological principle.

Before concluding our discussion of the cosmological principle, we may note an interesting work by Raine (1975) who, adopting a new formulation of Mach's principle, has shown that the principle requires that a strictly homogeneous perfect-fluid universe would be completely isotropic. However this cannot be looked upon as a derivation of the cosmological principle from the Mach principle as the homogeneity is introduced as an assumption. How far the perfect fluid assumption is crucial for the derivation also needs to be clarified.

1.3. THE ROBERTSON-WALKER METRIC

In pre-relativity physics the cosmological principle would simply mean a homogeneous distribution of matter and some restrictions on their velocity field. However for theories which picture the universe as a four-dimensional manifold and link physics with the geometry of this manifold, the assumptions of homogeneity and isotropy go much further - the entire geometry of space time must allow a group of motions involving six parameters such that all the points and directions in the three-space sections are equivalent. (However the converse, i.e. if the space time admits the group corresponding to spatial homogeneity and isotropy and we have a geometric theory of gravitation, the cosmological principle will hold good, is not, in general, true. A contrary example is provided in Brans-Dicke theory by McIntosh (1973) where with the Robertson-Walker line element one has a spatially non-uniform density and pressure. The contradiction arises essentially due to the occurrence of the scalar field which may have a lower symmetry than the geometry.)

Spatial homogeneity requires that the space time must admit a group of motions with three independent space like generators. The orbits of this group define a family of geodesically parallel space like hypersurfaces and the orthogonals to these hypersurfaces are time like. One may then write the line element in the form: (Eisenhart 1933; Taub 1951)

$$ds^2 = dt^2 + g_{ik}\, dx^i\, dx^k = dt^2 + d\sigma^2 \;, \tag{1.1}$$

with

$$d\sigma^2 = g_{ik}\, dx^i\, dx^k,$$

where g_{ik} is negative definite and there exist three independent Killing vectors.

If we further introduce the assumption of spatial isotropy it follows that the three spaces must be of constant curvature and have the metric:

$$d\sigma^2 = -\frac{R^2}{(1+kr^2/4)^2}\,(dr^2 + r^2\, d\theta^2 + r^2 \sin^2\theta\, d\phi^2) \;, \tag{1.2}$$

where R is a function of t alone and k=0, 1, or -1 and $0 \le \theta \le \pi$ and $0 \le \phi \le 2\pi$. The line element as given by (1.1) and (1.2) is known as the Robertson-Walker line element and it is important to note that it has been obtained simply by the application of group theoretic considerations to a 3+1 Riemannian manifold without the use of any field equations whatsoever (Robertson 1929, 1935, 1936; Walker 1936).

It is now usual to identify the t lines (here introduced as the geodetics orthogonal to the homogeneous varieties) with the world lines of matter (i.e. the galaxies). Implicitly it involves a number of assumptions as the following:

(i) The world lines of matter form a coherent pencil of geodesics, i.e. the motion is not turbulent and there is no non-gravitational interaction which, if present, would make the world lines non-geodetic. In the literature this assumption is referred to as the Weyl postulate (Weyl 1923, 1930).

(ii) The world lines are hypersurface orthogonal. This according to Gödel (1949) corresponds to irrotational motion, i.e. the matter does not rotate relative to the compass of inertia. This assumption is intimately connected with the Mach principle.

(iii) The frame of reference defined by the isotropy postulate coincides with the rest frame of cosmic matter. The condition of isotropy selects out a preferred frame of reference - an observer moving with a velocity relative to this frame will obviously notice an anisotropy. We are now assuming that the observers stationed on the galaxies will not observe any anisotropy. In a sense we may say that the translational velocity of the galaxies also vanish.

(iv) The velocity field is shear free. The meaning of this assumption will be more clear later on.

The Robertson-Walker line element leaves the function R and the parameter k undetermined. To link them up with the distribution of matter and radiation in the universe, one requires a theory of gravitation. We may note the very great demand made on the theory of gravitation - it must be correct on the one hand over distances as large as the linear dimensions of the universe, which is infinite for k=0 and -1 and on the other hand must be good also for almost arbitrary small distances and high densities if the 'big bang' (i.e. a state of infinite density) did actually exist in the history of the universe. The theory must also be valid over the entire time span of the universe.

2
NEWTONIAN GRAVITATION AND COSMOLOGY

2.1. FORMULATION AND CRITIQUE OF NEWTONIAN COSMOLOGY

It is difficult to say whether any of the present-day theories of gravitation would stand the very stringent requirements that we have set in the concluding lines of the last chapter. To be sure, the Newtonian law of gravitation is not one that can be accepted today. Yet, owing to the historic importance as well as the closeness between the results of the Newtonian law and the general theory of relativity, it seems worthwhile to have a look at Newtonian cosmology.

Newton's law of gravitation originated from Kepler's laws of planetary motion (and was not just a brainwave initiated by a falling apple). If one assumed the planetary orbits to be circular (which is indeed true to a very close approximation) the acceleration of the planets must be v^2/r and when this is combined with the third law of Kepler ($T^2 \propto r^3$) one obtains from Newton's laws of motion a force varying as $1/r^2$. However, Newton's enunciation of the law was striking as he realized that this interaction was not peculiar to the sun and the planets (or for the matter limited to the astronomical field) but held between all material particles. With the very exhaustive investigations of Laplace about a century later, the planetary motions, in all their finer peculiarities, could be explained satisfactorily on the basis of Newton's law but there remained the well known anomaly in the motion of mercury - an unaccounted perihelion motion of 43" per century.

The law of gravitation, as given by Newton was essentially an action at a distance formula, but following Laplace and Poisson we can present it in a field form: $\nabla^2\phi = -4\pi G\rho$. However this indicates instantaneous action. On the one hand the velocity of propagation being infinite, the field equations do not lead to any gravitational wave-phenomena and on the other hand an infinite velocity is repugnant to the ideas of the special theory of relativity and indeed

Poisson's equation is not Lorentz covariant.

It is thus clear that the Newtonian law, quite apart from any possible failure to give agreement with observations, cannot be an acceptable picture of gravitational interaction. Yet, as we shall see, the Newtonian law leads to results in striking agreement with those of relativistic cosmology, although the logical consistency of the Newtonian formulation is somewhat open to question (Layzer 1954; McCrea 1954; Heckman and Schucking 1959).

The Newtonian inverse-square law follows from the Poisson equation provided the potential vanishes sufficiently rapidly at infinity. While this may be a legitimate assumption for isolated distributions of matter in an otherwise empty space, it cannot be justified in the cosmological case where the distribution of matter extends over all space. One can thus anticipate that there would be divergent results in cosmology according to how one's considerations on the Poisson equation or the inverse-square law of Newton is based. This, apparently, is the explanation of the very different results obtained by Heckman and Schucking on one hand and Narlikar and later Davidson and Evans on the other, to which we shall return a little later.

McCrea and Milne (1934) showed that one could from Newtonian ideas, obtain in cosmology, equations which are formally identical to those of relativistic cosmology. At first sight the cosmological principle seems to be irreconcilable with the ideas of Newtonian mechanics. For, if all galaxies are on the same footing, then if the rest system of one be an inertial frame, so should the rest systems of all the others. Apparently, therefore, the only motion possible is one of uniform translation in which gravitation seems puzzlingly absent. However, with a slightly weaker conception of the cosmological principle, one can have accelerated motions.

Suppose that the particular point O, whose rest system is inertial, observes the velocity field $\underline{v}(\underline{r})$ for other particles, $\underline{r}$ being the position vectors of these particles. Then $\underline{v}'(\underline{r}-\underline{a})=\underline{v}(\underline{r}) - \underline{v}(\underline{a})$ gives the velocity field relative to the point $(\underline{a})$. Now cosmological principle is held to mean that both must observe exactly similar velocity fields, i.e. $\underline{v}' \equiv \underline{v}$, or

$$\underline{v}(\underline{r} - \underline{a}) = \underline{v}(\underline{r}) - \underline{v}(\underline{a}),$$

which leads to a linear relation between $\underline{v}$ and $\underline{r}$:

$$\underline{V}_i = \alpha_{ik}\, x_k \, . \tag{2.1}$$

(For a recent discussion of cosmological principle in Newtonian theory and a class of models in which the relation between $\underline{v}$ and $\underline{r}$ is not quite linear, see Evans and Davidson 1973.)

The motion is irrotational if $\underline{\nabla} \times \underline{v} = 0$ which means that α_{ik} is a symmetric matrix. We can then diagonalize α_{ik} by a co-ordinate transformation. If further, as is usually held, cosmological principle includes isotropy of the velocity field, the α_{ik} matrix must be a multiple of the unit matrix. We then get

$$v_i = f x_i \tag{2.2}$$

where f is a function of time alone. The volume dilation or expansion θ is $\underline{\nabla}.\underline{v} = 3f$. Let us introduce a new variable R by the relation:

$$\theta = 3f \equiv \frac{3\dot{R}}{R} \, . \tag{2.3}$$

We now use the hydrodynamical equations of Euler, the equation of continuity, and Poisson's equation for the gravitational field:

$$\frac{\partial v_i}{\partial t} + v_k \frac{\partial v_i}{\partial x_k} + \frac{1}{\rho}\frac{\partial p}{\partial x_i} = F_i = -\frac{\partial \phi}{\partial x_i} \tag{2.4}$$

$$\frac{\partial \rho}{\partial t} + (\rho v_i)_{,i} = 0 \tag{2.5}$$

$$F_{i,i} = -\phi_{,ii} = -4\pi\rho \qquad (2.6)$$

where in the last two equations we have used a comma followed by an index to indicate partial derivative with respect to the corresponding co-ordinate; also a summation is to be understood over any repeated index. We shall follow these conventions throughout the book. The units have been chosen as to make the Newtonian gravitational constant G equal to unity.

From equations (2.4) - (2.6), using the definition (2.3) and remembering that homogeneity requires that the pressure p and the density ρ are functions of time alone, we get

$$\rho R^3 = \text{const.} \qquad (2.7)$$

$$\frac{\ddot{R}}{R} = -\frac{4\pi\rho}{3} \qquad (2.8)$$

$$\frac{\dot{R}^2}{R^2} + \frac{K}{R^2} = \frac{8\pi\rho}{3} \quad . \qquad (2.9)$$

In the above and in all following discussion a superior point will indicate partial differentiation with respect to t. Equation (2.7) permits us to regard R as a measure of the linear dimensions of the system and equation (2.9) is essentially an energy equation k being the conserved total energy. For $k > 0$, it follows from equations (1.7) and (1.9) that an increasing R will attain a maximum and then decrease, i.e. the particles cannot escape from their mutual gravitational attraction. Again for $k < 0$, there will be no such maximum and the particles have energy enough to escape and the system proceeds monotonically to infinite dispersion. The case $k = 0$ corresponds to the situation where the particles have just enough energy for escape. As we shall see, general relativity gives exactly similar equations for isotropic universes, k and R being there the curvature parameter and the time-dependent scale function occurring in the Robertson-Walker line element respectively.

However, one may ask some very pertinent questions: firstly the deduction, as given above, seems justified only for the observer O whose rest system is inertial so that the Euler equations need not be modified by the addition of inertial forces. Would not the mathematical equations for the other observers and consequently the picture observed by them be essentially different? The answer is that the relation (2.2) obviously holds good for all observers with the same f and hence the same time-dependence of R. Mathematically, although the equations will involve an inertial-force term, as this term is independent of spatial co-ordinates, its divergence vanishes and hence equations (1.7) - (1.9) are not affected.

Secondly, is it permissible to use Newtonian ideas in such a case? Should we not expect that, in view of isotropy, the force field, which is a vector field with a preferred direction, vanishes everywhere? Further, the density distribution being homogeneous, why should there be a non-vanishing resultant force?

The answer is linked with the question of inertial forces. Each observer may consider himself at the centre of a spherically symmetric mass-distribution and, from Gauss' theorem, arrive at the result that the force on the particle at r is in the direction towards him and of magnitude $4\pi r\rho/3$. Thus there is a difference between the gravitational forces as reckoned by two observers of magnitude $4\pi\rho(r-r')/3$. However, in view of (2.8), this is exactly accounted for by the inertial force due to relative acceleration. Thus, on the cosmological level, there is no difference between the observations made by different observers.

2.2. A COMPARISON BETWEEN NEWTONIAN AND RELATIVISTIC COSMOLOGY

The agreement between the Newtonian formulae and the equations of general relativity has been sometimes attributed to the special condition of spherical symmetry. Thus Bondi (1947) showed that with the spherically symmetric line element

$$ds^2 = dt^2 - X^2\, dr^2 - Y^2(d\theta^2 \sin\theta^2\, d\phi^2) ,$$

where X and Y are functions of r and t and units are so chosen that the velocity of light is unity, one obtains for a dust distribution at rest in the above co-ordinate system the following equation in general relativity

$$\frac{\dot{Y}^2}{2} - \frac{1}{Y}\int_0^Y 4\pi Y^2 \rho \, dY = \frac{W^2 - 1}{2} \tag{2.10}$$

where W is a function of r alone (i.e. independent of t). Equation (2.10) is similar to equation (2.9) and can be interpreted as the analogue of the Newtonian energy-equation. Thus it has been argued that, subject to spherical symmetry, the motion would be similar to that in a Newtonian system. However the agreement between the Newtonian and general relativistic formulae goes much deeper than spherical symmetry and the cosmological principle, as has been demonstrated by the later investigations of Heckman and Schucking (1955), Raychaudhuri (1957), and Ellis (1971).

Let us drop equation (2.2) which is a consequence of the cosmological principle. We get, on taking the divergence of equation (2.4) and eliminating $\nabla^2\phi$ with the help of equation (2.5)

$$\frac{\ddot{R}}{R} = -\frac{4\pi\rho}{3} - \frac{1}{3}\sigma_{ik}\sigma_{ik} + \frac{1}{3}v_{[i,k]}v_{[i,k]} - \frac{1}{3}\left(\int \frac{dp}{\rho}\right)_{,ii}, \tag{2.11}$$

where R is now defined by

$$\theta \equiv v_{i,i} \equiv \frac{3\dot{R}}{R} \tag{2.12}$$

the superior point now signifying the operator

$$\frac{d}{dt} \equiv \frac{\partial}{\partial t} + v_i \frac{\partial}{\partial x_i} .$$

Again taking the curl of equation (2.4), we get

$$\frac{1}{R^2}\frac{d}{dt}(R^2 v_{[i,l]}) + \sigma_{ik} v_{[k,l]} + \sigma_{kl} v_{[i,k]} = 0 . \tag{2.13}$$

In the above equations the tensor $v_{i,k}$ has been split up as follows

$$v_{i,k} = v_{[i,k]} + \frac{1}{3}\theta\delta_{ik} + \sigma_{ik} , \qquad (2.14)$$

the square brackets denoting antisymmetrization. $v_{[i,k]}$ is thus the vorticity tensor and σ_{ik}, the shear tensor, is the trace-free part of the symmetric tensor $v_{(i,k)}$:

$$\sigma_{ik} = v_{(i,k)} - \frac{1}{3}\theta\,\delta_{ik} . \qquad (2.15)$$

Equation (2.13) gives the well known result that if the vorticity vanishes once for an element of the fluid, it will vanish always. Equations exactly similar to (2.11) and (2.13) appear in relativistic cosmology.

In a somewhat different manner, the similarity between the equations of Newtonian and relativistic cosmology has been exhibited in detail by Ellis (1971). If one splits up the symmetric tensor $\phi_{,ik}$ where ϕ is the Newtonian gravitational-potential into its trace $\phi_{,ii}$ and its trace free part $\phi_{i,k} - \frac{1}{3}\phi_{,ll}\delta_{ik}$, then Ellis shows that these play a role in Newtonian theory somewhat analogous to the Ricci tensor and Weyl tensor respectively in general relativity. However, while the Riemann - Christoffel tensor has 20 independent components, the tensor $\phi_{,ik}$ has only six. Thus the gravitational phenomena may be expected to be essentially more complicated and varied in any geometric theory of gravitation than in a Newtonian theory. Again a number of workers have considered it natural to assume the gravitational field in Newtonian theory to be spherically symmetric if the distribution of matter be homogeneous (even though there may be rotation) so that the trace-free parts of the tensor $\phi_{,ik}$ vanish. However the vanishing of the Weyl tensor would make the metric conformally pseudo-Euclidean which is by no means a consequence of density uniformity.

Put in another way, we may say that the total number of equations that one has to deal with in general relativity is larger except when reduced by symmetry conditions, and so the constraints to which the relativist is subject are much more stringent. It thus turns out that any relativistic model has a Newtonian analogue but the converse is not in general true. Indeed it is known, for example, that in general relativity shear-free, expanding, rotating, homogeneous solutions do not exist, but it is easy to construct such solutions in the Newtonian case as has been exhibited by Heckmann and Schücking (1955). Similarly, while the Gödel (1949) solution is the only stationary homogeneous solution in relativistic cosmology, the Newtonian scheme yields a host of such models when one modifies the Poisson equation by the introduction of a cosmological term (cf. Neumann 1896 and Seeliger 1896):

$$\phi_{,ii} = -4\pi\rho + \Lambda, \tag{2.16}$$

where Λ is a constant.

Another point is perhaps worth emphasizing. The introduction of geometry in general relativity brings in some features for which it is difficult to find analogues in Newtonian physics. Thus the case of an empty universe seems trivial and extremely artificial in Newtonian theory. However in general relativity we have a number of empty, non-trivial solutions of considerable interest, for example the Taub-Misner (Taub 1951; Misner and Taub 1969) and the Kasner (1927) universes. It is again difficult to find Newtonian analogues of geodetic incompleteness, closed time-like lines indicating a breakdown of causality, and topological peculiarities - such ideas have of late become topics of considerable interest in relativistic cosmology.

2.3. CAN COLLAPSE BE HALTED BY VORTICITY?

Equation (2.8) shows that a non-empty universe is necessarily non-static and further there is no minimum of R. Indeed with $\dot{R} < 0$, an approach to the state $R = 0$ is progressively accelerated. Thus either in the finite past or in the finite future (according as $\dot{R}$ is positive or negative at present) there is a collapse to a state of vanishing R and infinite density (see equation 2.7). It is of importance to know whether one can stop the development of this singular state by the introduction of vorticity, the possibility being suggested by equation (2.11). Conflicting answers to this question have been given by Heckmann (1961), Narlikar (1963), and Davidson and Evans (1971).

Heckmann considers shear-free expansion, thus $\sigma_{ik} = 0$. For the homogeneous universe the pressure gradient vanishes so that equations (2.11) and (2.13) yield

$$\frac{\dot{R}^2}{2} = + \frac{4\pi}{3} \int \rho R \, dR - \frac{1}{3} \frac{Q^2}{R^2} - k \, , \tag{2.17}$$

where Q and k are constants, Q being related to the rotational velocity ω:

$$\omega = \frac{Q}{R^2} \, ,$$

and $-k$ is the energy constant. If the cosmic material is in the form of a pressureless dust, equations (2.5) and (2.7) hold good and equation (2.17) gives

$$\frac{\dot{R}^2}{2} = \frac{M}{R} - \frac{1}{3} \frac{Q^2}{R^2} - k \quad . \tag{2.18}$$

The term in Q, i.e. that involving vorticity, would then be the controlling term for small R and there would be a minimum of R. However it is a little more realistic to consider the cosmic material to be a mixture of dust and radiation. One would then have

$$\rho = \rho_m + \rho_r \, ,$$

where ρ_m and ρ_r are the densities of dust and radiation respectively. If the dust and radiation are uncoupled, then for adiabatic expansion we have $\rho_m R^3$ = constant and $\rho_r R^4$ = constant. The r.h.s. of equation (2.18) would now contain two terms involving R^{-2} - one due to radiation and the other due to vorticity. In order that the singularity may be avoided one must have the inequality

$$\frac{4\pi\rho_r}{3} < \frac{1}{3} \frac{Q^2}{R^4} = \frac{1}{3} \omega^2$$

satisfied. At present with $\rho_r \sim 10^{-34}$ g cm^{-3} corresponding to the observed microwave background-radiation, this yields $\omega > 2 \times 10^{-4}$ seconds per century. This value, although apparently quite small, is nevertheless much greater than the upper limits to vorticity set by the isotropy of the background radiation (Hawking 1969).

Narlikar, however, has come to the conclusion that under any circumstance whatsoever, the Newtonian equations do not allow a singularity-free transition from a contracting to an expanding phase.

Equation (2.1) may be integrated in the form

$$r_i = a_{ik}\, r_k^0 ,$$

where the constants r_k^0 may be looked upon as the co-ordinates at a particular instant and the coefficients a_{ik} are functions of time alone.

Narlikar now takes the gravitational intensity as $\underline{F} = -4\pi\rho/3\ \underline{r}$ as would follow in case one had a finite spherically symmetric distribution of uniform density ρ. Using this expression in the Euler equation he was able to deduce the following equation

$$\Delta^2 \frac{d^4\Delta}{dT^4} + 7\Delta \frac{d^2\Delta}{dT^2} - 4\left(\frac{d\Delta}{dT}\right)^2 + 9\Delta = 0 , \qquad (2.19)$$

where $\Delta = \det|a_{ik}|$ and $T = (4\pi\rho_0/3)^{\frac{1}{2}}t$. A singularity is indicated by a vanishing of Δ and in particular if Δ is of rank two, one may say that the collapse is to a 2-surface, while for rank one the collapse is to a line, and for rank zero it is to a point.

Although it was not possible to integrate equation (2.19) analytically, Narlikar claimed from a study of the equation that a singularity $\Delta = 0$ could not be avoided. This was interpreted as indicating that the centrifugal force being normal to the axis of rotation may prevent a collapse into the axis of rotation, but a collapse along the axis is unaffected by rotation. In the Narlikar scheme, therefore, a rotation is invariably associated with shear (i.e. anisotropic velocity field) contrary to the situation discussed by Heckman and Schucking. The re-examination of the problem by

Davidson and Evans has considerably clarified the whole situation. They basically follow the same procedure as Narlikar, noting that, while Heckmann and Schucking use the Poisson equation, they do not use the expression $\underline{F} = -4\pi\rho\underline{r}/3$ for the gravitational intensity but rather fix up the force by the requirement of vanishing shear. Thus essentially the difference arises from the non-uniqueness of the solution of Poisson's equation in case of an infinite mass-distribution. Davidson and Evans feel that homogeneity leads naturally to spherical symmetry of $\underline{F}$ in Newtonian cosmology and this with the Gauss theorem allows them to write $\underline{F} = -4\pi\rho\underline{r}/3$. They find that with this form of the gravitational intensity, shear-free expansion with non-vanishing vorticity, as considered by Heckmann and Schucking, is not possible. Nevertheless their final results are different from those of Narlikar and this is attributed to some errors in Narlikar's considerations. Considering an axially symmetric system, Davidson and Evans found that there are both singularity free solutions and solutions which have a 'disc singularity', the disc being normal to the rotation axis, which is also the symmetry axis. In particular near the disc singularity $\Delta \sim t$, whereas for isotropic universes $\Delta \sim t^2$.

3
GENERAL RELATIVITY AND RELATIVISTIC COSMOLOGY

3.1. BASIC EQUATIONS OF ISOTROPIC RELATIVISTIC COSMOLOGY

Einstein was led to the general theory of relativity not in a search for a more accurate law of gravitation than Newton's so as to account for the perihelion motion of mercury or such other small effects, but instead by the ideas of general covariance (which, by the way, is a much stronger requirement than Lorentz covariance of special relativity) and the principle of equivalence (i.e. that the laws of nature would be the same in an inertial system as in a freely falling system). We shall not pause here to make a critical review of these principles but merely note that these do not lead uniquely to the Einstein field equations. There was for example the Nordstrom theory of a conformally flat geometry with the energy-momentum scalar $T = kR$. However Nordstrom's equations do not lead to the equations of motion (Guth 1969). To get the field equations Einstein imposed the additional requirement that the conservation principle for energy momentum must follow directly from the field equations. It turns out that one can then obtain the equations of motion from the field equations themselves.

The equations finally adopted were:

$$R_{\mu\nu} - \tfrac{1}{2} R g_{\mu\nu} = -k\, T_{\mu\nu} \qquad (3.1a)$$

or,

$$R_{\mu\nu} = -k(T_{\mu\nu} - \tfrac{1}{2} T g_{\mu\nu}) \qquad (3.1b)$$

where T is the energy-stress-momentum tensor, k a coupling constant (which with $G = c = 1$ is equal to 8π) and $R_{\mu\nu}$ is the Ricci tensor given by

$$R_{\mu\nu} = \Gamma^{\alpha}_{\mu\beta}\, \nabla^{\beta}_{\alpha\nu} - \Gamma^{\alpha}_{\mu\nu}\, \Gamma^{\beta}_{\alpha\beta} + \Gamma^{\alpha}_{\mu\alpha,\nu} - \Gamma^{\alpha}_{\mu\nu,\alpha} \qquad (3.2)$$

The $\Gamma^{\alpha}_{\mu\nu}$ are the Christoffel symbols defined by

$$\Gamma^{\alpha}_{\mu\nu} = \tfrac{1}{2}\, g^{\alpha\sigma}[g_{\sigma\mu,\nu} + g_{\sigma\nu,\mu} - g_{\mu\nu,\sigma}]\ . \qquad (3.3)$$

A more elegant procedure is to obtain the field equations from a variational principle with a suitable Lagrangian, a procedure first used by Hilbert in this case (for an interesting and illuminating account of the contribution of Hilbert in arriving at the field equations the reader may consult Mehra 1974). Since then the variational technique has been almost invariably used in the deduction of field equations.

This is not the place to give an account of the so-called crucial tests of general relativity or of the many different observational tests which have been proposed and are in different stages of experimentation. Although in some later chapters we shall have occasion to refer to some findings which have been claimed to contradict the predictions of general relativity, none of these findings may be said to have convinced the scientific community so far. In short we shall regard the general theory of relativity as the presently accepted theory of gravitation.

In the standard form of relativistic cosmology discussions are based on the cosmological principle and hence one accepts the Robertson - Walker line element which we have presented in Chapter 1. To use the field equations one must have the expression of energy-stress tensor. Assuming the cosmic matter to be a perfect fluid, we write:

$$T_{\mu\nu} = (p+\rho)v_\mu v_\nu - pg_{\mu\nu} \tag{3.4}$$

where p is the isotropic fluid-pressure and ρ the proper density of matter in the universe. Equations (3.1) give only two non-trivial relations:

$$\frac{8\pi\rho}{3} = \frac{R}{R^2} + \frac{\dot{R}^2}{R^2}\,, \tag{3.5}$$

$$-\frac{8\pi p}{3} = \frac{k}{R^2} + \frac{\dot{R}^2}{R^2} + \frac{2\ddot{R}}{R}\,. \tag{3.6}$$

In the next section we shall discuss the early developments in relativistic cosmology. Most of these are today merely of historical interest.

3.2. THE COSMOLOGICAL TERM AND THE STATIC-UNIVERSE MODELS

A look at the above equations shows that for a completely static universe (i.e. $\dot{R} = \ddot{R} = 0$), we must have $p = -\rho/3$. If, therefore, we insist on a non-negative pressure associated with a positive density of matter, the conclusion is unavoidable that the universe must be in a non-static condition.

Although this conclusion seems natural today, unfortunately even the greatest scientists seem incapable of freeing themselves from some deep-rooted bias, and thus Einstein, instead of making a prediction of spectral shift of light from distant galaxies, decided to modify the field equations (3.1) to

$$R_{\mu\nu} - \tfrac{1}{2} Rg_{\mu\nu} = -\Lambda g_{\mu\nu} - 8\pi T_{\mu\nu} \; . \tag{3.7}$$

The new term involving the constant Λ is called the cosmological term. It does not disturb the covariant character of the field equations or the conservation relation $T^{\mu\nu}{}_{;\nu} = 0$. The only restriction on Λ seemed to be that it must be small enough so as not to affect appreciably the planetary motions as these seemed to be explained quite satisfactorily by the original equations (3.1). With the introduction of this term it was possible to obtain static universes.

However, the later observations of red shift of light from distant galaxies found a ready explanation in non-static models of the universe and there remained no imperative need of the cosmological term. [For an alternative explanation of red shift as due to photon-photon interaction and criticisms of this hypothesis, see Pecker, Roberts, and Vigier (1972), Pecker, Tait and Vigier (1973), Cohen and Wertheim (1973), Geller and Peebles (1972), Puget and Schatzman (1974), Woodward and Yourgrau (1973), Aldrovandi, Caser, and Omnes (1973), Baum 1976, and Solheim, Barnes, and Smith (1976).] And indeed Einstein himself disclaimed this term later as 'the greatest mistake of my life' (quoted by Ruffini and Wheeler 1971). However one can find equally emphatic statements in favour of retention of the cosmological term - thus Eddington (1933) writes 'Λ-term is the strongest pillar of the theory of relativity and I would as soon think of reverting to Newtonian theory as of dropping the cosmical constant'. The arguments in favour of the cosmological term have ranged all the way from the philosophical to those linked with the clustering of quasar red-shifts, apparently negative values of the deceleration parameter, or considerations of elementary-particle physics. We shall enter into some of these in later chapters. Zel'dovich (1968) has shown that quantum fluctuations in vacuum may give rise to a term of the form $\Lambda g_{\mu\nu}$ in the stress-energy tensor. (For observational limit to Λ, see e.g. Campussano, Heidmann and Nieto, 1975.)

The present author, on his part, would prefer to wait for definitive observational evidence either this way or that and, pending that, maintain an open mind. Thus, for the most part, we shall work with the original field equations (3.1) but shall occasionally refer to the consequence of retaining the Λ-term.

Returning to the cosmological equations, we find that with (3.7) equations (3.5) and (3.6) are modified to

$$\frac{8\pi\rho}{3} = \frac{k}{R^2} + \frac{\dot{R}^2}{R^2} - \frac{\Lambda}{3} \tag{3.8}$$

$$-8\pi p = \frac{k}{R^2} + \frac{\dot{R}^2}{R^2} + \frac{2\ddot{R}}{R} - \Lambda, \tag{3.9}$$

so that for a static universe,

$$8\pi(p+\rho) = \frac{2k}{R^2}\,. \tag{3.10}$$

Thus k must be +1. Also,

$$8\pi(p+\rho/3) = 2\Lambda/3\,, \tag{3.11}$$

and hence Λ must also be positive.

At the present epoch, radiation constitutes only a very small fraction of the total energy ($\sim 10^{-4}$ say) and the random velocities of the galaxies are negligible compared to the velocity of light so that the pressure p may be neglected in comparison with the energy density ρ. We then get

$$4\pi\rho = \frac{1}{R^2} = \Lambda ,$$

and the line element now is

$$ds^2 = -\frac{R^2}{(1+r^2/4)^2}(dr^2+r^2d\theta^2+r^2\sin^2\theta\, d\phi^2) + dt^2 . \quad (3.12)$$

We may transform (3.12) to the form (Tolman 1934b)

$$ds^2 = -(dz_1^2+dz_2^2+dz_3^2+dz_4^2) + dt^2 , \quad (3.13)$$

with

$$z_1^2 + z_2^2 + z_3^2 + z_4^2 = R^2 . \quad (3.14)$$

Thus the metric is now of a flat space with 4 space-like and 1 time-like dimension and the space part of (3.12) is a hyperspherical surface in the hypothetical 4 space of (3.13). Such embedding of Riemann spaces in pseudo-Euclidean spaces of higher dimensionality may always be done - in the most general case a Riemann space of n dimensions will require a (pseudo-) Euclidean space of $n(n+1)/2$ dimensions. The difference between the dimensions of the embedding Euclidean space and the Riemann space is referred to as the class of the Riemann space. Thus the class of the isotropic cosmological space is unity. However, contrary to the speculations of some older cosmologists, no very important significance can be attached to this embedding so far as physics is concerned.

The solution (3.12) is referred to as the Einstein static universe. The Einstein space is closed in the sense that the proper distance $\int_0^\infty R\, dr(1+r^2/4)^{-1}$ converges but is nevertheless without any boundary. The total spatial volume is also finite and a light ray may come back to the point from which it started after circumnavigating the universe in a finite time (Tolman 1935). We may note in passing that even if we drop the cosmological principle but retain the Weyl postulate, the only static system is the Einstein universe (provided of course there is no pressure gradient or other non-gravitational forces) (Raychaudhuri 1955a).

We shall note another solution with non-vanishing Λ. One obtains with $p = \rho = 0$ (empty space):

$$ds^2 = dt^2 - e^{\alpha t}(dr^2 + r^2 d\theta^2 + r^2 \sin^2\theta \, d\phi^2) \ . \tag{3.15}$$

The line element, in this form, is not static but is said to be stationary as the physical conditions at different epochs do not change in any observable aspect. In fact the time dependence can be altogether eliminated with a transformation (Robertson 1928)

$$ds^2 = \left(1 - \frac{r^2}{R^2}\right) dt^2 - \frac{dr^2}{(1 - r^2/R^2)} - r^2(d\theta^2 + \sin^2\theta \, d\phi^2) \ . \tag{3.16}$$

The solution is known as the de Sitter universe and is characterized by a complete absence of matter in the conventional outlook. However in the late forties the de Sitter line element in form (3.15) acquired a new interest from the point of view of the steady-state theory (Bondi and Gold 1948; Hoyle 1948). This is the only line element which satisfies the perfect cosmological principle (i.e. that besides spatial homogeneity and isotropy, the overall picture of the universe must not change with time as well) and at the same time allows for an expansion (red shift). If the field equations were augmented by the introduction of a creation term (Hoyle 1948) or if negative pressures are considered admissible (McCrea 1951) the line element can be reconciled with the presence of matter in finite density.

3.3. THE ENERGY-MOMENTUM TENSOR AND BULK VISCOSITY

We have assumed so far that the energy-momentum tensor of the cosmic matter is that characteristic of a perfect fluid. Let us look at the situation from a different angle. The cosmological principle in the form of the Robertson - Walker line element puts severe restrictions on the nature of the T^{μ}_{ν} tensor; for using the field equations, all the terms T^{α}_{β} for $\alpha \neq \beta$ vanish and $T^1_1 = T^2_2 = T^3_3$ i.e. the normal stresses are equal and the tangential stresses as well as the energy flux vanishes. One cannot thus have any shear viscosity or heat conduction. Nor can one have any electromagnetic or gravitational radiation except in the diffuse form. However the line element does allow a bulk-viscosity term.

The bulk viscosity may be shown to vanish both for non-relativistic and ultrarelativistic gases but in general it does not vanish in the intermediate region. What is of more interest, is that bulk viscosity need not vanish for a mixture of ultrarelativistic and non-relativistic particles. Thus it may not vanish at some stages of the universe. For the Robertson - Walker line element Weinberg (1971) obtains the following general form for the $T_{\mu\nu}$ components:

$$T_{ij} = \left(p - 3\zeta \frac{\dot{R}}{R}\right) g_{ij}$$

$$T_{00} = \rho, \qquad T_{io} = 0$$

where i,j run from 1 to 3 (i.e. over the space co-ordinates and 0 stands for the time co-ordinate.) ζ is the coefficient of bulk viscosity and when it arises from some massless particles (e.g. photons, neutrinos, or gravitons) Weinberg finds

$$\zeta = 4bT^4\tau\left(\frac{1}{3} - \frac{\partial p}{\partial \rho}\right)^2$$

where b is the Stefan constant for the case of photons and gravitons and is 7/8th times that constant in case of neutrinos τ is the mean free-time for the massless particles, and T the temperature.

The influence of bulk viscosity on the temporal behaviour of isotropic universes have been studied by Treciokas and Ellis (1971), Neugebauer and Strobel (1969), and Nightingale (1973). It has been shown that the cycles of expansion and contraction for the closed universe (forgetting for a moment the singular state - see the following sections) become asymmetrical and the maximum world radius tends to larger and larger values. Nightingale has found that there is also a slight increase in the time scale and this may have important influence on the abundance of lighter nuclei if they were produced in the early universe.

Singularity free isotropic cosmological solutions have been constructed by Murphy (1973), Heller, Klimek, and Suszycki (1973), and Heller and Suszycki (1974). However in all these works the bulk viscosity is assumed to have special forms - $\zeta = \alpha\rho$ in Murphy's work and ζ = constant in the papers of Heller *et al.* - inconsistent with the Weinberg formula and violate the Hawking-Penrose energy condition (see Chapter 8 and also Das and Agarwal 1974; Roy and Singh 1974; Lukács 1976).

3.4. THE SINGULARITY OF ISOTROPIC MODELS

Let us go back to the discussion of equations (3.5) and (3.6). It is clear that they do not determine the sign of $\dot{R}$ and this is to be expected owing to the time-reversal symmetry that obtains in the general theory of relativity. However we get from these equations

$$\frac{\ddot{R}}{R} = -\frac{4\pi}{3}(\rho+3p) , \tag{3.17}$$

so that with positive pressure and density $\ddot{R} < 0$ and thus R cannot have any minimum and,either in the finite past or in the finite future (according as $\dot{R}$ is at present positive or negative), one necessarily has a singular state corresponding to $R = 0$. That this signularity is physical follows from the simple fact that from the conservation relation $d(\rho R^3) + p d(R^3) = 0$, ρR^3 increases monotonically or remains constant as R decreases and hence $\rho \to \infty$ as R goes to zero. This singularity of infinite density is the so called 'big bang'.

We shall return to the question of singularity of more general cosmological models in Chapter 8.

3.5. TIME BEHAVIOUR OF ISOTROPIC MODELS

The function R determines the time behaviour of models. However the two equations (3.5) and (3.6) (or (3.8) and (3.9) if one retains the cosmological constant) involve three functions of time p, ρ, and R. One has therefore to supplement the two equations by some other condition - usually an equation of state connecting p and ρ. At the present epoch the condition that seems most appropriate is $p = 0$ while at early hot stages of the universe the relation might have been $p = \rho/3$ corresponding to an ultrarelativistic gas and radiation. However, we shall not at first impose any definite relation between p and ρ but simply keep in mind the usual physical requirements $p \geq 0$, $\rho > 0$ and following Tolman (1934b) discuss the various possibilities that may arise. We shall take the case of closed space ($k = +1$) and open space ($k = 0, -1$) separately.

For $k = +1$, equation (3.8) gives

$$\dot{R}^2 = \frac{8\pi}{3}\rho R^2 + \Lambda\frac{R^2}{3} - 1 \,. \tag{3.18}$$

Writing

$$Q \equiv \frac{3}{R^2} - 8\pi\rho \,, \tag{3.19}$$

we get

$$\frac{\mathrm{d}Q}{\mathrm{d}R} = -\frac{6}{R^3} + 24\pi\frac{(p+\rho)}{R} \,. \tag{3.20}$$

In obtaining the last equation we have used the conservation relation

$$\dot{\rho} = -3\,(p+\rho)\,\frac{\dot{R}}{R} \tag{3.21}$$

which is simply a consequence of the field equations. For an extremum of Q, we get from equations (3.19) and (3.20)

$$Q_0 = \frac{1}{R^2} + 8\pi p > 0 \,. \tag{3.22}$$

As the condition for maximum (<), inflexion (=) and minimum (>) of Q we obtain after a second differentiation:

$$\dot{p} \lesseqgtr (p+\rho)\,\frac{\dot{R}}{R} \,. \tag{3.23}$$

Let us study the plot of Q against R. With $p > 0$, ρR^3 is a decreasing function of R remaining constant in the limiting case $p = 0$. Hence from equation (3.19) as $R \to 0$,

$Q \to -\infty$, and for $R \to \infty$, $Q \to +0$ as $3/R^2$. For some intermediate values, Q must therefore have positive non-vanishing values. Hence there must be at least one maximum of Q as shown by B in Fig.3.1. The portions marked a, b, c are not definitely

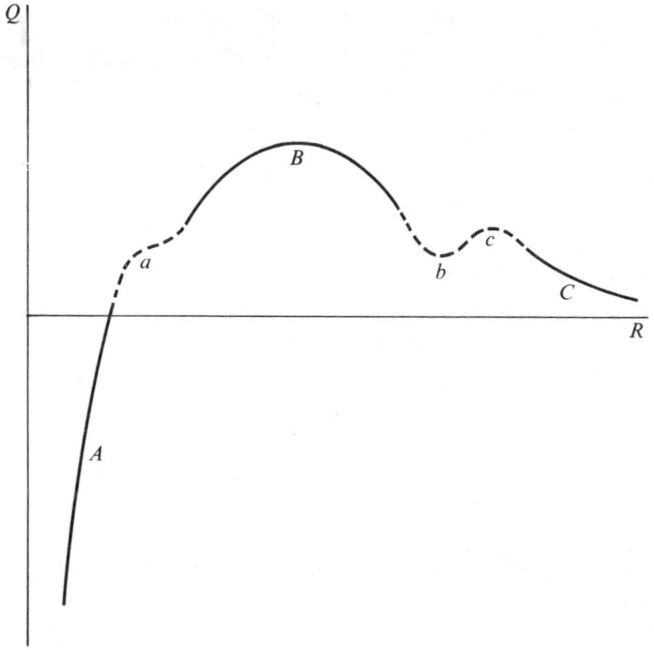

Fig.3.1. Plot of $Q \equiv 3R^{-2} - 8\pi\rho$ against R. (Tolman, 1934b).

known but are meant to indicate the possibilities of additional extrema of Q. Using equations (3.18) and (3.19) we can write

$$\frac{\dot{R}}{R} = \pm\left(\frac{\Lambda - Q}{3}\right)^{\frac{1}{2}}$$

so that we are restricted to values of $Q \leq \Lambda$. Further $Q = \Lambda$ corresponds to a maximum, inflexion, or minimum of R according as $\mathrm{d}Q/\mathrm{d}R$ is positive, zero, or negative. The following cases arising for different values of Λ are of interest.

(i) If $\Lambda > \Lambda_E$ where Λ_E denotes the highest maximum of Q, then Q is nowhere equal to Λ and we have a monotone expansion or contraction with R running over all values from zero to infinity. These are known as Lemaitre universes and at the highest value of Q, $\dot{R}$ will have a minimum. The value of $\dot{R}$ at this minimum decreases as Λ approaches Λ_E so that the universe may spend a considerable time near about this state - which, with a small enough value of $\dot{R}$, may be considered a quasi-static state. This feature of the Lemaitre universe has been sometimes considered to be of importance in the formation of galaxies and in providing an explanation for the bunching of quasar red-shifts (Petrosian, Salpeter, and Szekeres 1967; Shklovski 1967; Brecher and Silk 1969). These models also give a longer time-scale than those with $\Lambda = 0$.

(ii) If $\Lambda = \Lambda_E$, there is a singular solution corresponding to the Einstein static universe. The other possible solutions either asymptotically start from or tend towards the Einstein state, all the derivatives of R tending to vanish as $Q \to \Lambda$. These models starting from the Einstein state were at one time quite popular and considerable efforts were made to show that a perturbation of the unstable Einstein universe would make it expanding (Lemaitre 1931; Sen 1933).

(iii) For $0 < \Lambda < \Lambda_E$, there are at least two values of R for which $Q = \Lambda$. In this case either R may increase from zero to a finite maximum and then turn back or alternatively R may run between a finite minimum and infinity. Should there be additional maxima and minima in the Q - R curve (as at a, b, c) the possibility exists of a regular oscillation between two finite volumes. However a minimum of Q would require, in view of equation (3.23) an increase of pressure with decreasing density.

(iv) For $\Lambda \leq 0$, there occurs only one value of R for which $Q = \Lambda$ and the only possibility is one of expanding from a singular state of zero volume and turning back at a finite value of R to collapse again to $R = 0$.

A number of points regarding the above discussion may be emphasized - (a) the direction of time flow may be reversed so that all the $R - t$ curves may be described in either direction, (b) all the possibilities except that of regular oscillation are realizable for dust as well ($p=0$), (c) for $\Lambda = 0$, the possibility is unique whether p vanishes or not.

For the open models $k = -1$ or 0, equation (3.24) remains unchanged provided we redefine Q by the following relations

$$Q \equiv -8\pi\rho - \frac{3}{R^2} \quad \text{for} \quad k = -1$$

$$\equiv -8\pi\rho \quad\quad \text{for} \quad k = 0 \; .$$

We thus have Q everywhere negative and tending to $-\infty$ as $R \to 0$ and to zero as R tends to infinity. Hence for $\Lambda \geq 0$, there can only be monotone increase of R from zero to infinity while for $\Lambda < 0$, the universe runs between a singular state of vanishing volume and a finite maximum.

Besides this general discussion, we can integrate the equations in terms of simple functions in several cases. We mention below a few examples.

3.5.1. The Einstein - de Sitter universe.
In this case $\Lambda = p = k = 0$. We get $R = at^{2/3}$, $\rho = \rho_0 t^{-2}$. This case allows all integrations to be performed easily and has received the greatest attention from astronomers (Einstein and de Sitter 1932).

3.5.2. The oscillating dust universe.
Let $\Lambda = p = 0$, $k = +1$. We get

$$\dot{R} = \left(\frac{\alpha}{3R} - 1\right)^{\frac{1}{2}} ,$$

where the constant α is defined by $\rho R^3 = \alpha$. We may have the integral in a parametric form

$$R = \frac{\alpha}{6}(1 - \cos\psi)$$

$$t = \frac{\alpha}{6}(\psi - \sin\psi)$$

so that R apparently oscillates between 0 and $\alpha/3$ with a time period $\pi\alpha/3$. However $R = 0$ is a singular state and the field equations cannot describe the passage through this state so that it is not permissible to say that we have an oscillatory universe. Nevertheless the oscillatory model has considerable appeal as it somehow agrees with the naive thesis that the universe must be everlasting without any catastrophic beginning or end. (This solution is originally due to Friedmann 1922.)

3.5.3. The radiation universe

In this case $p = \rho/3$, $\Lambda = 0$, $k = +1$. One gets

$$\frac{8\pi\rho R^4}{3} = \beta \text{ (const.)}$$

$$R = (2\beta^{\frac{1}{2}}t - t^2)^{\frac{1}{2}}$$

so that R runs between 0 and $\beta^{\frac{1}{2}}$ and near the singularity $R \sim t^{\frac{1}{2}}$. This may be a good representation of some early stage of big-bang models (Tolman 1931).

3.5.4. The dust cum radiation universe

Tolman considered the case where the dust and radiation are non-interacting. The total density ρ is split up into two parts, ρ_m due to pressureless dust and ρ_r due to radiation with pressure $p = \rho_r/3$. We get

$$\rho = \rho_m + \rho_r$$

$$p = p_r = \rho_r/3$$

$$\rho_m R^3 = \text{const.} = \frac{3M}{4\pi}$$

$$\rho_r R^4 = \text{const.} = A$$

$$\dot{R} = \left(\frac{\Lambda R^2}{3} - k + \frac{2M}{R} + \frac{4\pi A}{3R^2}\right)^{\frac{1}{2}} .$$

For $k = +1$ and $\Lambda = \Lambda_E$ the value required for the Einstein static universe, the integral of the above equation has been given by Tolman (1934b), while for $\Lambda = 0$, the general solution was obtained by Alpher and Herman (1949). A particular solution for open models as given by Chernin (1965, 1968) reads

$$R = a(e^{\eta}-1)$$

$$t = ka(e^{\eta}-\eta-1)$$

where a and t are constants and η runs from zero to infinity (see also McIntosh 1967).

More recently Landsberg and Park (1975) have discussed the case where there is an interaction (i.e. an exchange of energy) between the gas and radiation.

3.6. GEOMETRY OF THE THREE SPACE

The three-space sections orthogonal to the t-lines which are the world lines of cosmic matter have the metric

$$d\sigma^2 = -\frac{R^2}{(1+kr^2/4)^2}[dr^2+r^2d\theta^2+r^2\sin^2\theta d\phi^2] .$$

This is of positive, negative, or zero curvature according as k is +1, -1, or 0. Further, owing to the factor R, the proper distance between any two points with constant coordinates is time dependent but the shape of figures remains unchanged as all lengths change in the same ratio (isotropic dilation without shear).

We can transform the three-space line element to

$$d\sigma^2 = -R^2(du_1^2+du_2^2+du_3^2+du_4^2)$$

where $u_1^2+(u_2^2+u_3^2+u_4^2)k = 1$, so that the three space is referred to as spherical, flat, of hyperbolic according as k is = 1,0, or -1. Thus, while the Einstein space is spherical, the de Sitter or the steady-state space is flat. In the case of a spherical space one may identify the antipodal points and then the space is called elliptic.

The line element $d\sigma^2$ is, except for any singularity of $R(t)$, regular over the entire region $0 \leq r \leq \infty$ for $k = 0$ and +1, but has an apparent singularity at $r = 2$ for $k = -1$. However this singularity occurs at an infinite proper distance and cannot be physical as all space points are equivalent under the group of motions. Obviously the scalars that can be constructed from the Riemann-Christoffel tensor would also not show any singularity.

One can calculate the total proper volume in different cases:

$$V = \int (-g)^{1/2} \, dr d\theta d\phi \; .$$

For the elliptic space, $V = \pi^2 R^3$. One says, therefore, that the space is closed and finite although unbounded. For $k = 0$ and -1, V diverges and the space is said to be open. (The descriptions open and closed are not quite unambiguous. Thus Geroch (1967) points out that if one considers the manifold

$$ds^2 = dt^2 - (1+t^2)^2 \, d\chi^2 - R^2(d\theta^2 + \sin^2\theta d\phi^2)$$

with R a constant, θ, ϕ spherical polar co-ordinates, and χ an angular co-ordinate with $0 \leq \chi \leq 2\pi$, the space-like sections t = constant are compact but the space-like sections $t = \frac{1}{2}\tan(\chi + t)$ are non-compact.)

There has been some bias in favour of compact spaces in so far as they dispose of the questions of boundary conditions and are sometimes considered to be a pre-requisite for the satisfaction of Mach's principle (Wheeler 1964b). Such considerations in favour of closed spaces seem to date back to Einstein (1950). However it should not be forgotten that Einstein along with de Sitter (1932) advocated a model with flat space. The present author's feeling is that in such questions philosophical considerations can hardly be expected to settle the issue and one must wait for the verdict from observations. Unfortunately, so far the observational data are by no means decisive. However we shall consider the observational situation later.

3.7. THE AGE OF THE UNIVERSE

The time that has lapsed from the singular state to the present state of the universe, commonly called the age of the universe, is of considerable interest. For $\Lambda = 0$, we have from equation (3.17) that $\ddot{R}/R$ is always negative so that the age would be less than $R/\dot{R}$.

We shall presently see that $\dot{R}/R$ is closely linked with the red shift of light from distant galaxies, which was first observed by Hubble. $\dot{R}/R$ is thus called the Hubble's parameter and is usually denoted by H, so that we have for a non-empty universe with $\Lambda = 0$, the age $\tau < H^{-1}$.

We may be a little more specific if we take $p = 0$. Then eliminating ρ from equation (3.5) with the help of the conservation relation ρR^3 = constant, we get

$$\dot{R}^2 R - \dot{R}_0^2 R_0 = -k(R - R_0),$$

where the subscript 0 refers to any specified state, conveniently called the present state. We have

$$\dot{R} = \left[\frac{\dot{R}_0^2 R_0}{R} - k\left(1 - \frac{R_0}{R}\right)\right]^{\frac{1}{2}},$$

and for the time interval from the singular state $R = 0$ to the present state is

$$\tau_0 = \int_0^{R_0} \left[\frac{\dot{R}_0^2 R_0}{R} - k\left(1 - \frac{R_0}{R}\right)\right]^{-\frac{1}{2}} \mathrm{d}R$$

and according as $k = 1, 0$ or -1

$$\tau_0 < \tfrac{2}{3} H_0^{-1} \quad \text{if} \quad k = +1$$

$$= \tfrac{2}{3} H_0^{-1} \quad \text{if} \quad k = 0$$

$$> \frac{2}{3} H_0^{-1} \quad \text{but} \quad < H_0^{-1} \text{ if } k = -1 \; .$$

In section 11 we shall give formulae connecting the age of the universe with H_0 and the deceleration parameter q_0. Ideas about the age of the universe or rather bounds to it may be had from a variety of considerations such as the ages of stars and elements in the galaxy ($\sim 8 \times 10^9$ years), the age of globular clusters (16×10^9 years), and nucleochronometry based on plausible models ($11 - 18 \times 10^9$ years) (Hainebach and Schramm 1976). (See also Beek, Hilf and Mair, 1973). These may set severe constraints on q_0 if H_0 be known. However there does exist considerable uncertainty about the value of H_0. Besides the age may be increased by introducing the cosmological term. Indeed for the Lemaitre point source models one can almost arbitrarily increase the 'coasting period' and correspondingly the time to the singularity, i.e. the age of the universe.

3.8. PATHS OF PARTICLES AND LIGHT RAYS AND HORIZONS

The t lines themselves are geodesics and correspond to the fact that cosmic matter is at rest in the co-ordinate system being used. It is easy to see that the expansion of the universe tends to make all test bodies conform to the general cosmic motion. Thus the geodesic equation

$$\frac{\mathrm{d}^2 t}{\mathrm{d}s^2} + \Gamma^0_{\mu\nu} \frac{\mathrm{d}x^\mu}{\mathrm{d}s} \frac{\mathrm{d}x^\nu}{\mathrm{d}s} = 0$$

gives

$$\left(\frac{\mathrm{d}t}{\mathrm{d}s}\right)^2 - 1 = AR^{-2}; \quad A \text{ being a constant.}$$

But for the time-like lines, $(\mathrm{d}s/\mathrm{d}t)^2 = 1 - u^2$ where u is the co-ordinate velocity. Hence

$$1 - u^2 = (1 + AR^{-2})^{-1}$$

showing that u decreases as R increases, i.e. any arbitrary test particle ultimately tends to share the general cosmic motion in an expanding universe. However for a tachyon whose world line is space-like, the co-ordinate velocity in the cosmic frame increases and becomes arbitrarily large in the open models at a finite time (Raychaudhuri 1974; Narlikar and Sudarshan 1976).

As, owing to the symmetry, we are free to choose our origin, any light ray may be considered to be travelling radially. For a light ray leaving r_1 at t_1 and reaching r_2 at t_2

$$\int_{r_1}^{r_2} \frac{\mathrm{d}r}{(1+kr^2/4)} = \int_{t_1}^{t_2} \frac{\mathrm{d}t}{R} \quad .$$

There arise some interesting situations if the r.h.s. integral converges either as $t_2 \to \infty$ or $t_1 \to 0$. In the first case the light ray cannot go beyond a certain value of the co-ordinate radius in its whole career, while in the latter case, at any finite time any particular observer has received light signals only from up to a limited co-ordinate distance. These two situations have been described as existence of event horizon and particle horizon respectively by Rindler (1956). Thus he defines them in the following manner: an event horizon for any given fundamental observer A is an hypersurface in space time which divides all events into two non-empty classes: those that have been, are or will be observable to A and those that are forever outside A's possible powers of observation. A particle horizon for any given fundamental observer A and cosmic instant t_0 is a surface in the instantaneous space $t = t_0$ which divides all fundamental observers into two non-empty classes: those that have already been observable to A (and vice versa) and those that have not.

For the Robertson-Walker line element, we have already seen that with $\Lambda = 0$, the open universes have a monotone expansion to $R \to \infty$ as $t \to \infty$. If we take $R \sim t^{\alpha}$, then from equation (3.17) we get

$$\frac{4\pi}{3}(\rho+3p) = -\alpha(\alpha-1)t^{-2} \quad , \tag{3.24}$$

showing that for non-negative values of $(\rho+3p)$, we must have

$$0 \le \alpha \le 1 \ . \tag{3.25}$$

With this restriction on α, $\int^{\infty} R^{-1}dt$ obviously diverges and there does not exist any event horizon.

In case of closed universes ($k = +1$), with $p = 0$, it turns out that the time for light to circumnavigate is exactly equal to the period of a complete cycle.

Again for both the open and closed models, if $\Lambda = 0$, R tends to zero at a finite time in the past for expanding models and this instant is conveniently taken to be $t = 0$. With $R \sim t^{\alpha}$ as $t \to 0$, we again have the relation (3.25) and so $\int_0 R^{-1}dt$ converges at the lower limit irrespective of any particular equation of state and we have a particle horizon. We have already seen that the particle horizon gives rise to difficulty in understanding the uniformity of the universe, especially as revealed by the isotropy of the background radiation. It appears that if the particle horizon is to be removed one must have (a) an anisotropic universe as suggested by Misner, but as we have already seen in Chapter 1 Misner's expectations have not proved correct, (b) a metric not satisfying the Einstein equations, as proposed by Hoyle and Narlikar (1970), or (c) a negative pressure $p \le -\rho/3$. Such a relation appears physically unacceptable but one may recall McCrea's suggestion that $p = -\rho$ to make the steady-state models consistent with the equations of general relativity. Whittaker (1966) again considered $p = -2/3\ \rho$ in a different context.

Eliminating p between equations (3.24) and the conservation relation

$$\dot{\rho} = -3(p+\rho)\dot{R}/R$$

one obtains

$$\rho = \frac{3\alpha^2}{8\pi t^2} + \frac{C}{t^{2\alpha}} \ ,$$

where C is an integration constant. Owing to the condition

(3.25) the first term on the r.h.s. will be dominating for $t \to 0$ and one gets

$$\rho = \frac{3\alpha^2}{8\pi t^2}, \quad p = -\left(1 - \frac{2}{3\alpha}\right)\rho$$

for sufficiently early epochs. Thus if p lies between 0 and ρ (the Zel'dovich limit, see Zel'dovich 1962), we get

$$\frac{1}{3} \leq \alpha \leq \frac{2}{3},$$

the bounds on α being now further narrowed down compared to (3.25).

3.9. THE RED SHIFT

The light emitted at t_i from the co-ordinate position r_i reaches the origin at t_0 where

$$\int_{t_i}^{t_0} \frac{dt}{R} = -\int_{r_i}^{0} \frac{dr}{(1+kr^2/4)} .$$

Again the light emitted at $t_i+\delta t_i$ reaches at $t_0+\delta t_0$ where

$$\frac{\delta t_i}{R(t_i)} = \frac{\delta t_0}{R(t_0)} .$$

Thus for non-static models $\delta t_i \neq \delta t_0$. If δt_i is the time period of emission, δt_0 will be that of reception and as the co-ordinate times also measure the proper times, the wave lengths λ_i and λ_0 of emission and reception will be related by

$$\frac{\lambda_i}{R(t_i)} = \frac{\lambda_0}{R(t_0)}$$

or, writing $z = (\lambda_0-\lambda_i)/\lambda_i$,

$$z = \frac{R(t_0)}{R(t_i)} - 1 . \qquad (3.26)$$

Observation shows a shift towards larger wave lengths, so

z is positive and R must be an increasing function of time at the present epoch and one is thus led to the idea that the galaxies are receding from each other - in short one has an expanding universe.

On the basis of the formula (3.26) one would expect that the shift, as at present observed, would be uniquely determined by the epoch at which light left the source, i.e. by the radial position of the source. (Of course there will be peculiar velocities but for galaxies these are small compared to the velocity of light and so should not affect significantly the shift except in case of nearby galaxies.)

In recent years, a number of observers have apparently found marked difference in the red shift for objects they had reason to believe to be at identical distance (Arp 1970, 1971; Bottinclli and Gougienheim 1973; Jaakola 1973; Collin-Souffrin, Pecker, and Tovmassion 1974; de Vaucouleurs and de Vaucouleurs 1973; Burke and Hartwick 1974; Tifft 1972; Harrison 1975). At thc IAU Symposium No.58 at Canberra, Arp as well as Balkowski, Bottinelli, Chamaraux, Gougienheim, and Heidmann (1974) were of the opinion that observations favoured the existence of anomalous red-shifts, but a contrary opinion was expressed by Sargent (1974). There was a discussion of the case of the Stephan's quintet and in the opinion of Longair (1974) the data are not irreconcilable with a simple Hubble-shift. An analysis by Jaakkola (1971) showed apparently that E, So, and Sa galaxies have excessive negative and Sb and Sc galaxies have excessive positive red-shifts. As yet there does not seem convincing evidences to disbelieve that the red shifts are essentially of the Hubble type (see the discussion in Field 1973); however should these peculiar red-shifts be substantiated they may well lead to profound changes not only in cosmology but also in physics in general (Cowan 1968; Hoyle and Narlikar 1971, 1972).

Returning to equation (3.26) in cases where the source and the observer are near enough, we get

$$z = \frac{\dot{R}}{R}\,\mathrm{d}t$$

where $dt = t_0 - t_i$ is the time of light travel. With the velocity of light as unity $dt = r$, the distance between the source and the observer, so that

$$z = \frac{\dot{R}}{R} r \equiv Hr \ .$$

This is the famous linear relation between z and r for small distances and H, as we have already mentioned, is called the Hubble constant. In general H would be a function of the epoch of observation. The linear relation has been well confirmed over a fairly wide range, although for nearby galaxies (distances 5 to 25 Mpc) there are apparently striking departures from both linearity and isotropy but this is usually interpreted as due to differential rotation and expansion of the supercluster to which the nearby galaxies are supposed to belong. The possibility of gravitational waves contributing a steady red-shift has been investigated by Dautcourt (1975).

3.10. RELATIONS INVOLVING H, z, AND THE DECELERATION PARAMETER q

We shall now present some useful and simple relations, they are not essentially new but simply the field equations rewritten in terms of the Hubble constant H, the red shift z, and the deceleration parameter q. The last one is defined by the relation

$$q \equiv \frac{\ddot{R}/R}{(\dot{R}/R)^2} \tag{3.27}$$

and is thus a pure number.

Equations (3.5) and (3.6) read

$$\frac{8\pi\rho}{3} = \frac{k}{R^2} + H^2 \tag{3.28}$$

$$-\ 8\pi p = \frac{k}{R^2} + H^2(1-2q) \ , \tag{3.29}$$

and adding

$$8\pi(p+\rho/3) = 2qH^2 \tag{3.30}$$

so that q must be ≥ 0, the sign of equality corresponding to an empty universe. For a pressure-free dust universe, equation (3.29) shows that the curvature is positive, zero, or negative according as q is greater than, equal to, or less than $\frac{1}{2}$, respectively. In particular excluding the trivial case of empty universe ($q = 0$) only for the case $k = 0$, would q be a constant. With $p = 0$, eliminating ρ and k from equation (3.28) with the help of equations (3.29) and (3.30) we get, remembering that ρR^3 and k are constants

$$2q_0 H_0^2 \left(\frac{R_0}{R}\right)^3 = H_0^2 \left(\frac{R_0}{R}\right)^2 (2q_0-1) + H^2$$

where the subscript zero indicates the values at a particular instant, usually identified with the present. Using equation (3.25) we have finally

$$H^2 = H_0^2 (1+z)^2(1+2q_0 z) \tag{3.31}$$

a relation which we shall have occasion to use in our later discussion.

3.11. RELATION BETWEEN LOOK-BACK TIME AND z AND THE AGE OF THE UNIVERSE AND H,q

We have from the definition of H and equation (3.26), for a given epoch of observation, the relation between the emission time t and z

$$dt = \frac{1}{H}\frac{dR}{R} = -\frac{1}{H}\frac{dz}{1+z}\,. \tag{3.32}$$

Substituting from equation (3.31) in equation (3.32), we get

$$dt = -\frac{1}{H_0}\frac{dz}{(1+z)^2(1+2q_0 z)^{\frac{1}{2}}}\,. \tag{3.33}$$

The integration of equation (3.33) yields the relation between t and z, i.e. the epoch of emission of light and the red shift (or equivalently the relation between t and R

as $R = R_0(1+z)^{-1}$). In particular as $z \to \infty$, $R \to 0$ and the corresponding value of t gives the age of the universe.

Case I. $q = q_0 = 0$, corresponding to the empty universe, ($k = -1$); we get

$$t = H_0^{-1}[1 - (1+z)^{-1}]$$

and for the age of the universe $\tau = H_0^{-1}$.

Case II. $q = q_0 = \frac{1}{2}$, $k = 0$. This is the case of the Einstein-de Sitter universe.

$$t = \frac{2}{3} H_0^{-1} [1 - (1+z)^{-3/2}]$$

and the age of the universe $\tau = 2/3\ H_0^{-1}$.

Case III. $0 < q < \frac{1}{2}$ (open universe with $k = -1$)

$$t = 2H_0^{-1} q_0\alpha^{-1} [(2q_0)^{-1} - \beta^{-\frac{1}{2}}(1-\alpha\beta^{-1}) + \tfrac{1}{2}\alpha^{-\frac{1}{2}} \log\left\{\frac{\alpha^{-\frac{1}{2}}+\beta^{-\frac{1}{2}}}{\alpha^{-\frac{1}{2}}-\beta^{-\frac{1}{2}}} \times \frac{\alpha^{-\frac{1}{2}}-1}{\alpha^{-\frac{1}{2}}+1}\right\}]$$

where $\alpha \equiv 1-2q_0$, $\beta \equiv 1+2q_0 z$ and the age of the universe

$$\tau = (H_0\alpha)^{-1}[1 + q_0\alpha^{-\frac{1}{2}}\log \frac{\alpha^{-\frac{1}{2}}+1}{\alpha^{-\frac{1}{2}}-1}] .$$

Case IV. $q > \frac{1}{2}$ (closed universe, $k = +1$).

$$t = (H_0\alpha)^{-1}\{\beta^{\frac{1}{2}}(1+z)^{-1} -1 +2q_0\alpha^{-\frac{1}{2}}[\tan^{-1}\alpha^{\frac{1}{2}} - \tan^{-1}(\beta^{-\frac{1}{2}}\alpha^{\frac{1}{2}})]\}$$

where $\alpha \equiv 2q_0-1$ and $\beta = 1+2q_0 z$. The age of the universe is

$$\tau = (H_0\alpha)^{-1}[-1 + 2q_0\alpha^{-\frac{1}{2}}\tan^{-1}\alpha^{\frac{1}{2}}] .$$

The look-back time to $z = 3$ (believed for some time to be the limiting value of quasar red-shift), is

$$t_{z=3} = 0{\cdot}51 \times H_0^{-1} = 1{\cdot}0 \times 10^{10} \text{ years},$$

where we have used $H_0 = 50$ Km. $s^{-1}Mpc^{-1}$.
The occurrence of the inverse circular functions in this case corresponds to the fact that for closed universes R undergoes a cyclic change (forgetting for the moment the difficulty associated with the singular state $R = 0$). The length of the cycle, i.e. the interval between two successive singular states of $z \to \infty$ is given by

$$T = 2\pi\, H_0^{-1}\, \alpha^{-3/2}\, q_0$$

which with the old value of $q_0 = 1$, becomes

$$T = 2H_0^{-1}\pi = 12 \times 10^{10} \text{ years.}$$

More recently Sandage (1975) and Sandage and Tammann (1975a,b) have arrived at much smaller values of q_0 and now vanishingly small or even negative values of q_0 are being suggested from analysis of observational data (Gunn and Oke 1975) (see however, Kruszewski and Semeniuk 1976). There is also lack of unanimity about the value of H_0 (de Vaucouleurs 1972) but in general the majority seem to favour a value $H_0 = 50$ km s^{-1} Mpc^{-1}. We shall return to these problems when we discuss the analysis of observational data. (The above value of H_0 has been given by Lang, Lord, Johanson, and Savage (1975) also, from a composite Hubble diagram of 663 normal galaxies, 235 radio galaxies, and 265 quasi-stellar objects.)

4
THE ANALYSIS OF OBSERVATIONAL DATA

4.1. THE QUESTIONS TO BE SETTLED

An analysis of the observational data may be expected to give answers to a variety of questions, typical of which are the following:

(i) Is the universe on a large scale really homogeneous and isotropic and the Robertson-Walker line element a good representation of the geometry of the universe? If so what exactly is this large scale? If not, what is the precise nature and magnitude of the departures from homogeneity and isotropy - in particular what are the here-now values of the shear (i.e. departure of the expansion from isotropy) and the vorticity in the motion of cosmic matter?

(ii) If the Robertson-Walker metric is good for the universe, what is the form of the function R and the value of the parameter k? Short of a complete knowledge of R, one would like to know the present values of the Hubble constant H and the deceleration parameter q.

(iii) What are the here-now values of p and ρ? What are the relative contributions of baryons, neutrino, and photons which go to build up the total energy density?

(iv) Are the equations of general relativity satisfied? What is the position of other field equations, say those of Brans and Dicke?

(v) What is the value of the cosmological constant? In particular is it zero?

(vi) As we are looking to distant galaxies, we are also looking backwards in time. What is the nature of the systematic evolutionary changes, if any, in the properties, e.g. absolute magnitude, spectra, dimension, etc. of the galaxies.

(vii) What were the relative concentrations of different nuclear species in the pre-galactic stage? The synthesis of heavy nuclei is believed to be taking place in stars but the stellar theory cannot account for the observed abundance of lighter elements like deuteron and helium. The early stage of the big-bang universe is a likely locale for the formation of light elements. Indeed, should there come some definitive observational evidence that deuterons were actually produced cosmologically, (cf. Ostriker and Tinsley 1975) big-bang models would get very impressive support.

However, as in any field, it is generally easy to pose questions, but much more difficult to answer them. Observational cosmology is perhaps worst in this respect. The presently available data are of such uncertain accuracy and in many cases there are so many obscure factors involved that it is nearly impossible to give unequivocal answers to most of the questions that we have raised.

The analysis that we present here assumes *a priori* that the Robertson-Walker line-element gives correctly the geometry of the universe. So far there does not exist any direct contradiction to this assumption - rather the observed high degree of isotropy of the microwave-radiation background lends powerful support to the idea that the universe has been isotropic at least over a considerable time in the past (i.e. up to the last scattering surface for the radiation).

The most important data are the observed relations between the apparent magnitudes (or more correctly the intensity of blackening on the photographic plates) and the red shift of spectral lines of the galaxies. One tries to translate this relation into one between the theoretically introduced quantities like the co-ordinate r, the curvature parameter k, and the function R (or rather the Hubble constant

H and the deceleration parameter q).

4.2. PROPER DISTANCE, LUMINOSITY DISTANCE, AND DISTANCE BY APPARENT SIZE

Distance in cosmology may be defined in a variety of ways. The most direct definition is that at time t, the distance of the point of co-ordinate r from the origin is given by

$$D_{\text{proper}} = R\int_0^r \frac{\mathrm{d}r}{(1+kr^2/4)}\,, \tag{4.1}$$

R having the value appropriate for time t. While this distance is invariant for purely spatial transformations as also transformations of the time scale, it is however operationally unrealizable as it involves the idea of a simultaneous measurement of all elements over the range 0 to r. Actually a beam of light, which forms our only link with distant galaxies, would meet different elements at different instants involving a variation of R.

The luminosity distance is defined as the distance which will preserve the validity of the inverse square law for the fall of intensity. If L_1 is the rate of emission of energy by a point source (the emission being assumed isotropic) (situated at co-ordinate r_1) per unit solid angle and l_1 the energy incident per unit time on unit area placed normally to the rays at the origin, then

$$l_1 = \frac{L_1}{R_0^2}\,\frac{(1+kr_1^2/4)^2}{r_1^2}\,\frac{1}{(1+z)^2}\;.$$

It is not difficult to understand the above relation - the light at the instant of reception is spread over a spherical surface of area

$$\frac{4\pi R_0^2 r_1^2}{(1+kr_1^2/4)^2}\,,$$

R_0 being the value of R at the instant of reception t_0. The red shift affects the intensity in two ways - each individual photon has its energy degraded by the factor $(1+z)^{-1}$ and the interval between the reception of photons is also increased by the factor $(1+z)$ - hence the $(1+z)^{-2}$ term on the r.h.s.

From the above formula the luminosity distance is defined as

$$D = \frac{R_0 r_1}{(1+kr_1^2/4)}(1+z) = \frac{R_0^2}{R_1}\frac{r_1}{(1+kr_1^2/4)} . \tag{4.2}$$

The above definition of luminosity distance is somewhat defective from the operational point of view for the following reasons (cf. Hoyle 1962): (i) while the source luminosity is spread over a wide spectrum, in actual observations only the radiation in some limited frequency range is studied. Again this corresponds to different frequency ranges of the source galaxies depending on their distances. (ii) Galaxies are not point sources, thus instead of the luminosity the surface brightness is measured.

Another way of defining the distance is by the apparent size of distant objects. The distance ξ so defined is given by

$$\xi = \frac{R_1 r_1}{(1+kr_1^2/4)} \tag{4.3}$$

where $R_1 = R(t_1)$, t_1 being the instant of emission of the light from the source at r_1, reaching the observer at the origin at t_0. ξ has also been sometimes called the 'corrected luminosity distance' (Kristian and Sachs 1966).

The relation between ξ and D is thus

$$D = \xi(1+z)^2 . \tag{4.4}$$

Of course, for near enough objects, $z \ll 1$ and all the three distances D_{proper}, D, and ξ approach each other.

4.3. THE RELATION BETWEEN THE LUMINOSITY DISTANCE, AND THE APPARENT MAGNITUDE

What the astronomer observes directly is a blackening of the photographic plate. He must first convert these blackenings to apparent magnitudes. (The magnitude is essentially a measure of the luminosity on a logarithmic scale; thus if l is the luminosity the magnitude m is given by

$$m = \text{const.} - 2{\cdot}5 \log l \tag{4.5a}$$

or,

$$l = A10^{-0{\cdot}4m} \tag{4.5b}$$

so that if the luminosity ratio is, say 100, the magnitude difference will be 5. It should be noted that a higher magnitude corresponds to lower luminosity.) The observed blackening requires correction due to (i) heterochromatic absorption in the earth's atmosphere, i.e. an absorption which is wave length dependent, (ii) the selective sensitivity of the photographic plate to different regions of the spectrum, and (iii) the effect of red shift on the spectrum which increases with distance.

If $E^*(t)$ is the energy (integrated over all frequencies) emitted by the source at time t per unit time, then the energy l^*_{bol} received per unit time per unit area outside the earth's atmosphere (i.e. before heterochromatic absorption) is

$$l^*_{\text{bol}} = \frac{E^*(t)}{4\pi D^2} \tag{4.6}$$

where D is the luminosity distance of the source. If m_{bol} is the apparent bolometric magnitude of the source (i.e. the magnitude taking into account all wave lengths) and M_{bol} is the absolute bolometric magnitude, then by definition

$$m_{\text{bol}} - M_{\text{bol}} = 2{\cdot}5 \log_{10} \left[\frac{L^*_{\text{bol}}}{l^*_{\text{bol}}}\right] \tag{4.7}$$

where L^*_{bol} is the bolometric luminosity if the source were at a standard distance of 10 parsecs. Hence, if D is in parsecs, we get from equations (4.6) and (4.7):

$$\log_{10} D = 0{\cdot}2\ (m_{\text{bol}} - M_{\text{bol}}) + 1\ . \tag{4.8}$$

This relation allows us to calculate D from m_{bol} if the absolute magnitude of the source M_{bol} be known. Now, in the nearer galaxies, one can identify Cepheid variables - stars whose luminosity varies - there being a definite relation between the period of variation and the luminosity. Thus from observations of this period and the apparent luminosity of these stars, one can determine their distances. In this way the distances of the nearby galaxies have been determined. Further, once their distances are determined, the observed apparent magnitudes lead to a determination of their absolute magnitudes. It has been found that the absolute magnitudes are close to an average value (~ -20) with a small scatter. (This is not quite true. In fact the astronomer has to observe members of galactic clusters and he takes say the 5th brightest member of the cluster. This is assumed to have a standard absolute magnitude.) Assuming now that this absolute magnitude is the same for all distant galaxies, we may determine their luminosity distances from observation of their apparent magnitudes. This simple method, of course, assumes implicitly that no appreciable evolutionary change has taken place in the absolute magnitudes of the galaxies for we are observing the distant galaxies at an earlier epoch. It is precisely to emphasize the possible evolutionary effect that we have taken E^* as a function of the time of emission t.

One can put, therefore,

$$M_{\text{bol}} = M'_{\text{bol}} - 2\cdot 5 \log \frac{E^*(t)}{E^*(t_0)} \tag{4.9}$$

where M'_{bol} is the value of the absolute magnitude arrived at from observations on nearer galaxies for which the instant of emission may be taken to be effectively the same as the instant of reception. It now follows

$$\log_{10} D = 0\cdot 2 \left(m_{\text{bol}} - M'_{\text{bol}} + 2\cdot 5 \log \frac{E^*(t)}{E^*(t_0)}\right) . \tag{4.10}$$

The above relation allows us to find D if the evolution of E^* be known.

Next we consider heterochromatic absorption, plate

selectivity, and red shift. Let $E(\lambda,t)\,d\lambda dt$ denote the energy emitted within the wavelength range λ to $\lambda+d\lambda$ in time dt at t. Then by the definition of luminosity distance, the energy incident per unit time per unit area is

$$\frac{E(\lambda_0/(1+z),t)\,d\lambda_0/(1+z)}{4\pi D^2}$$

where the factor $(1+z)$ has appeared due to the fact that the radiation received in the wave length range λ_0 to $\lambda_0+d\lambda_0$ were at the source in the region $\lambda_0/(1+z)$ to $(\lambda_0+d\lambda_0)/(1+z)$. The recorded energy is therefore

$$\frac{1}{1+z}\cdot S(\lambda_0)\cdot\frac{E(\lambda_0/(1+z),t)\,d\lambda_0}{4\pi D^2}$$

where $S(\lambda_0)$ is the characteristic function taking account of the wavelength-dependent absorption in the atmosphere and the sensitivity of the plate. The integrated intensity is therefore

$$= \frac{1}{4\pi(1+z)D^2}\int_0^\infty S(\lambda_0)E\left(\frac{\lambda_0}{1+z},t\right)d\lambda_0 \quad .$$

The bolometric intensity is obtained by putting $S(\lambda_0) = 1$ (i.e. the case of no absorption and ideally faithful recording of all wavelengths) and is

$$= \frac{1}{4\pi(1+z)D^2}\int_0^\infty E\left(\frac{\lambda 0}{1+z},t\right)d\lambda_0$$

$$= \frac{1}{4\pi D^2}\int_0^\infty E\left(\frac{\lambda_0}{1+z},t\right)\frac{d\lambda_0}{1+z}$$

$$= \frac{1}{4\pi D^2}E^*(t) \quad .$$

The value of the absolute magnitude also has to be corrected for this heterochromatic absorption and selective sensitivity. We have

$$m_{het} - M'_{het} = 2\cdot 5\ \log_{10}(L^{*\prime}_{het}/l_{het})$$

where the subscript 'het' indicates heterochromatic quantities. Thus

$$m_{\text{bol}} - M'_{\text{bol}} = m_{\text{het}} - M'_{\text{het}} - K \tag{4.11a}$$

where

$$K = 2{\cdot}5 \log_{10}\left\{\frac{(1+z)^{-1} \int_0^\infty E[\lambda_0/(1+z),t]\,d\lambda_0}{\int_0^\infty E(\lambda,t_0)\,d\lambda}\right\}$$

$$- 2{\cdot}5 \log_{10}\left\{\frac{(1+z)^{-1} \int_0^\infty S(\lambda_0)E[\lambda_0/(1+z),t]\,d\lambda_0}{\int_0^\infty S(\lambda_0)E(\lambda_0,t_0)\,d\lambda_0}\right.$$

$$= 2{\cdot}5 \log_{10} \frac{E^*(t)}{E^*(t_0)} - 2{\cdot}5 \log_{10}\left\{\frac{\int_0^\infty S(\lambda_0)E[\lambda_0/(1+z),t]\,d\lambda_0}{(1+z)\int_0^\infty S(\lambda_0)E(\lambda_0,t_0)\,d\lambda_0}\right\}.$$

In the above expression for K, the second term is specially difficult to evaluate as it requires a knowledge of the spectral distribution of the light from galaxies, which again involves the evolutionary changes as well as the nature of dependence of the atmospheric absorption and the photographic sensitivity on wavelength. Finally

$$\log_{10}D = 0{\cdot}2\,(m_{\text{het}} - M'_{\text{het}} - K_0) + 1 + \text{evolution correction} \tag{4.11b}$$

where K_0 is the part of K arising due to the function $S(\lambda)$ and red shift:

$$K_0 = 2{\cdot}5 \log_{10}(1+z) - 2{\cdot}5 \log_{10}\left\{\frac{\int_0^\infty S(\lambda_0)E[\lambda_0/(1+z)]\,d\lambda_0}{\int_0^\infty S(\lambda_0)E(\lambda_0)\,d\lambda_0}\right\}.$$

The function $S(\lambda)$ has to be evaluated purely empirically. Assuming for the moment that K_0 has been evaluated, the above equation determines D as a function of m_{het} subject, of course, to the evolution correction. But D itself is a function of z, r_1, R, and the curvature parameter k (equation 4.2). Our task in the next section will be to express

D explicitly as a function of z, H_0, and q_0 and thus to obtain a relation between m_{het} and z (both of which are observable) and H_0 and q_0 which we seek to determine from the observational data.

The effect of evolution may be considered in the manner of Sandage (1961b) and Solheim (1966). Thus we can take the effect of evolution to be a change ΔM_τ in the absolute magnitude,

$$\Delta M_\tau = 2\cdot5 \log \frac{E^*(t_0-\tau)}{E^*(t_0)}$$

where $t = t_0 - \tau$ is the instant of emission. The transmission time

$$\tau = \int_{R_0}^{R} \dot{R}^{-1}\, \mathrm{d}R .$$

One may assume that a galaxy contains two types of stars - the evolving giants and the white dwarfs. The luminosity of the latter group does not change appreciably with time. Sandage takes the law of evolution of the luminosity of giants in an E-galaxy to be inversely proportional to the 4/3rd power of time. Thus

$$\Delta M_\tau = 2\cdot5 \log[P_0 + (1-P_0)\left(1 - \frac{\tau}{t_0}\right)^{-4/3}]$$

where P_0 is the ratio E_D/E_{tot}, E_D being the energy emitted by dwarfs and E_{tot}, the total emission of the galaxy. Thus

$$E_{\text{tot}} = E_D + E_0 \left(\frac{t}{t_0}\right)^{-4/3}$$

E_0 being the energy emitted by giants at the present epoch $t = t_0$. τ/t_0 has been calculated by Refsdal, Stabell, and de Lange (1967) as a function of q_0, σ_0, and z ($\sigma_0 \equiv 4\pi p_0/3H_0^2$), so that one can compute the evolution correction ΔM_τ if one can determine P_0 which is in the range $0\cdot25 - 0\cdot5$.

Oke and Sandage (1968) and Oke (1971) from observations of giant elliptical galaxies up to $z = 0\cdot46$ concluded that the evolution correction is probably negligible up to these values of z. However Tinsley (1972) made a calculation based on the following two assumptions about the stellar population of giang ellipticals: (a) the stars are formed all in an initial burst of duration negligible compared to their present age, and (b) the initial mass function for stars of mass $\leq 1\cdot5$ M. is a power law. Tinsley arrived at a significant evolution correction and concluded that the apparent value of q_0 must be at least $0\cdot5$ greater than its true value (Tinsley 1973a). In any case the evolution correction is certainly not negligible (Baldwin, Danziger, Frogel, and Persson 1973) and may be of an extent as to reduce the value of q_0 by even $1\cdot2$ (Rose and Tinsley 1974).

The evolutionary correction from a population synthesis of ellipticals is (Gunn and Oke 1975; Gunn and Tinsley (1975)

$$\Delta M(z) \approx -(1\cdot3 - 0\cdot3x)\ln\left[\frac{t_0 - t(z)}{t_{gal}}\right]$$

where t_0 is the present age of the universe, $t(z)$ is the age at z, t_{gal} is the present age of the galaxy, and x is the slope of the initial-mass function. The evolution correction to the value of q_0 is then, to the first order,

$$\Delta q \equiv (q_0)_{apparent} - q_0 = \frac{2\cdot5}{\ln 10}\cdot\frac{2}{2-\alpha}\,\frac{1\cdot3-0\cdot3x}{H_0\,t_{gal}}$$

where $\alpha \approx 0\cdot7$ is the slope of the flux - diameter relation for ellipticals. Recent results on red- and infrared-line indices demand $x \leq 1$ if the initial-mass function is represented by a power law. Δq is thus essentially positive and taking $x = 1$, and $t_{gal} \approx t_0 \approx 1\cdot8 \times 10^{10}$ years, we have a lowest limit to the evolution correction

$$(\Delta q)_{min} \approx \frac{2\cdot5}{\ln 10}\cdot\frac{2}{1\cdot3}\quad.$$

With q_0 apparently nearly $\sim +1$, such a correction may make q_0 negative. However, the possibility of another evolutionary correction has been pointed out by Ostriker and

Tremaine (1975). This would arise from the possible accretion of dwarf companions by giant galaxies. This would be of opposite sign and may be of similar magnitude as the correction due to stellar evolution. Thus the final role of evolutionary correction remains obscure (Gunn and Tinsley 1976).

4.4. THE RELATION BETWEEN THE LUMINOSITY DISTANCE D, THE RED SHIFT z, THE HUBBLE CONSTANT H_0, AND THE DECELERATION PARAMETER q_0

To obtain a relation between D and z (the relation will also involve R in the form H_0 and q_0) two procedures are open. A power-series expansion may be used and, although this uses the Robertson - Walker line element, the field equations are not used. Alternatively one may try to obtain the relation between D and z in a closed form by using the field equations.

Let us consider the method of Taylor series expansion. Writing $t = t_0 - \tau$, we have

$$R(t) \equiv R(t_0-\tau) = R_0(1-\alpha_1\tau+ \frac{\alpha_2}{2}\tau^2 - \ldots) \tag{4.12}$$

where

$$\begin{aligned} \alpha_1 &\equiv (\dot{R}/R)_0 \equiv H_0 \\ \alpha_2 &\equiv (\ddot{R}/R)_0 \equiv -q_0H_0^2 , \end{aligned} \tag{4.13}$$

the subscript 0 indicating the values at $t = t_0$. We perform a term by term integration of both sides of the following equation by using the above expansion

$$\int_0^{r_1} \frac{dr}{(1+kx^2/4)} = \int_t^{t_0} \frac{dt}{R} \tag{4.14}$$

and obtain

$$\frac{1}{R_0}\,[\tau + \alpha_1 \frac{\tau^2}{2} + \frac{1}{6}\,(2\alpha_1^2 - \alpha_2)\tau^3 + \ldots\,]$$

$$= r_1 - k\,\frac{r_1^3}{12} + \frac{k^2 r_1^5}{80} \quad \ldots$$

so that solving for r_1

$$r_1 = \frac{\tau}{R_0}\,[1 + \frac{\alpha_1\tau}{2} + \frac{1}{6}\left(2\alpha_1^2 - \alpha_2 + \frac{k}{2}\,\frac{1}{R_0^2}\right)\tau^2 + \ldots\,]\ . \qquad (4.15)$$

Utilizing the relation (3.26) (p. 40) i.e.

$$z = \frac{R_0}{R} - 1$$

we get, using the series for $R(t)$

$$z = \alpha_1\tau + \left(\alpha_1^2 - \frac{\alpha_2}{2}\right)\tau^2 + \ldots$$

or,

$$\tau = \frac{z}{\alpha_1} + \frac{\frac{1}{2}\alpha_2 - \alpha_1^2}{\alpha_1^3}\,z^2 + \ldots \qquad (4.16)$$

so that substituting from equation (4.16) in (4.15)

$$r_1 = \frac{z}{\alpha_1 R_0}\left(1 + \frac{\alpha_2 - \alpha_1^2}{2\alpha_1^2}\,z\ \ldots\right)\ . \qquad (4.17)$$

Finally, using equation (4.17) in equation (4.2), we get for the luminosity distance

$$D = \frac{z}{\alpha_1}\left(1 + \frac{\alpha_2 + \alpha_1^2}{2\alpha_1^2}\,z\ \ldots\right)$$

or,

$$\log_{10} D = \log_{10}\left(\frac{z}{\alpha_1}\right) + \frac{1\cdot 086}{5}\left(1 + \frac{\alpha_2}{\alpha_1^2}\right) z + \ldots\ . \qquad (4.18)$$

Combining equation (4.18) with (4.11b), we get

$$m_{\text{het}} - k_0(z) = 5 \log_{10} z + 1{\cdot}086 \left(1 + \frac{\alpha_2}{\alpha_1^2} + \lambda^*\right) z$$
$$+ M'_{\text{het}} - 5 \log_{10}\alpha_1 - 5 \qquad (4.19)$$

where $K_0(z)$ is written to emphasize that K_0 is also a function of z and the term in λ^* arises from evolution, where, in addition, only the part linear in z is supposed to be retained.

It is noteworthy that the curvature parameter does not occur in the linear term in (4.19). Thus conceivably the evolution effects may be more important than those due to curvature in the m - z relation.

The merit of the series method is obvious - it does not introduce any assumptions regarding either the field equations or the state of cosmic matter (e.g. any equation of state). However, our complete lack of knowledge about the successive derivatives of R makes it impossible to determine how rapidly the series converges and how far we are justified in taking say the first or the first two terms and making estimates (Heckmann 1968; McVittie 1938).

Observationally the data are well represented by a linear relation between $\log z$ and corrected m (for z not very large):

$$\log z = 0{\cdot}2\, m + \text{const.} \qquad (4.20)$$

The fact that the relation besides being linear has the theoretically expected slope 0·2 is an impressive evidence in favour of the Robertson-Walker metric.

The constant in the ($\log z - m$) equation determines $\alpha_1 \equiv H_0$ (compare equations (4.19) and (4.20)). As we have already noted, it is now generally taken to have the value 50 km s^{-1} Mpc^{-1}. It should be noted that the determination of H_0 depends on the value we take for M'_{het}, the absolute magnitude derived from observations on near galaxies. During the past 25 years, there has been a successive upgrading in the evaluation of M'_{het} and consequently H_0 has gone down.

Even now, depending on the method of determining the

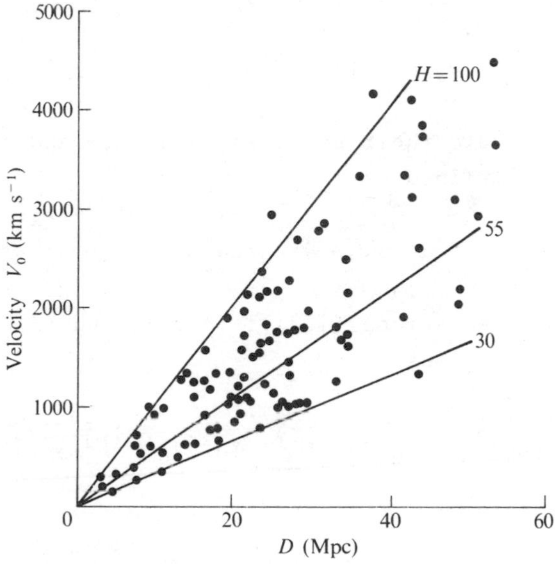

Fig.4.1. Plot of recessional velocity of galaxies V ($\equiv cz$) against the luminosity distance D. H_0 = 55 km s^{-1} Mpc^{-1} gives the best fit - all the points lie between the straight lines with slopes H_0 = 100 km s^{-1} Mpc^{-1} and 30 km s^{-1} Mpc^{-1}. (Sandage and Tammann, 1975a.)

distance scale, the estimated value of H_0 has varied, ranging all the way from 40^{+15}_{-13} km s^{-1} Mpc^{-1} to 110 ± 10 km s^{-1} Mpc^{-1} (Branch and Patchett 1973; de Vaucouleurs 1972). An estimate of H_0 without the determination of the distance scale has been made by Kirshner and Kwan (1974) and they give the value H_0 = 60 ± 15 km s^{-1} Mpc^{-1} (see also Rowan-Robinson 1976 ; Jaakkola and Le Denmot 1976; de Vaucouleurs 1977; Lynden-Bell, 1977, Bottinelli and Gougienheim, 1976).

The m - z relation can, in principle, determine $1 + \alpha_2/\alpha_1^2 + \lambda^*$ if the observations go to fairly high values of z and are of sufficient accuracy; but we have already seen the great uncertainty that hangs over the evolution of the factor $K_0(z)$ (Whitford 1971; Oke and Sandage 1968).

Let us now consider the m - z relation using the field equations of general relativity. In case $\Lambda \neq 0$ but p = 0, one can write from the equations (3.8) and (3.9) (p. 24):

$$\Lambda = -3(q_0-\sigma_0)H_0^2 \qquad (4.21)$$

$$k = -(1+q_0-3\sigma_0)H_0^2R^2 \qquad (4.22)$$

where q_0 and H_0 have their usual significance and the density parameter σ_0 is defined as

$$\sigma_0 \equiv 4\pi\rho_0/3H_0^2 \ . \qquad (4.23)$$

With the help of equations (4.21), (4.22), and (4.23), equation (3.8) (p.24) gives

$$\frac{\dot{R}}{R} = H_0[2\sigma_0 v^3 - (q_0-\sigma_0) + (1+q_0-3\sigma_0)v^2]^{\frac{1}{2}} \qquad (4.24)$$

where

$$v \equiv \frac{R_0}{R} = 1+z \ .$$

Using equation (4. 24) in (4.14), we get

$$\chi \equiv \int_0^{r_1} \frac{\mathrm{d}r}{(1+kr^2/4)} = \frac{1}{H_0R_0}\int_1^v [2\sigma_0 v^3-(q_0-\sigma_0)+(1+q_0-3\sigma_0)v^2]^{-\frac{1}{2}}\,\mathrm{d}v \qquad (4.25)$$

Equation (4.25) can be integrated in closed form in terms of elementary functions when Λ vanishes, i.e. $q_0 = \sigma_0$ (Mattig 1958). This we shall consider presently. An integration in terms of elementary functions is also possible for $\sigma_0 = 0$, $\Lambda \neq 0$ (de Sitter universe with the steady-state line element). For other cases, the elliptical integral has been numerically evaluated for various values of q_0 and σ_0 (Solheim 1966; Refsdal, Stabell, and de Lange 1967). Consequently one can also numerically compute the red-shift magnitude relations and thereby estimate the values of q_0 and σ_0. More recently m - z relations have been given in terms of Weirstrass *pe* functions for the cases $p = 0$ and $p = \rho/3$ for arbitrary values Λ and k (Kaufman 1971, 1973; Kaufman and Schucking 1971).

Returning to the case $\Lambda = 0$, we write the integral of equation (4.25) for various values of k

$$\chi = \begin{cases} \cos^{-1}\left(\dfrac{\beta}{1+z} - 1\right) - \cos^{-1}(\beta-1) & \text{for } k = +1 \\ (2\beta)^{\frac{1}{2}} - \left[\dfrac{2\beta}{(1+z)}\right]^{\frac{1}{2}} & \text{for } k = 0 \\ \cosh^{-1}(\beta+1) - \cosh^{-1}\left(\dfrac{\beta}{1+z} + 1\right) & \text{for } k = -1 \end{cases} \tag{4.26}$$

where

$$\beta = \begin{cases} (2q_0-1)/q_0 & k = +1 \\ \left(\sigma_0 \dfrac{H_0^2 R_0^2}{2}\right)^{-1} & k = 0 \\ (1 - 2q_0)/q_0 & k = -1\ . \end{cases} \tag{4.27}$$

(Note that q_0-1 vanishes for $k = 0$ and the relation $\beta = (\sigma_0 H_0^2 R_0^2/2)^{-1}$ also holds for $k = \pm 1$.) Again from equation (3.8) and the definition of χ (see 4.25), we get

$$D = R_0(1+z) \times \begin{cases} \sin\chi & \text{for } k = +1 \\ \chi & \text{for } k = 0 \\ \sinh\chi & \text{for } k = -1\ . \end{cases} \tag{4.28}$$

Finally combining equations (4.26), (4.27), and (4.28) we get

$$D = \frac{1}{H_0 q_0^2}\{q_0 z + (q_0-1)[(1+2q_0 z)^{\frac{1}{2}} - 1]\} \tag{4.29a}$$

for $q_0 > 0$ whatever be the value of k and for $q_0 = 0$, which corresponds to $\rho = 0$ and $k = -1$,

$$D = \frac{z}{H_0}\left(1 + \frac{z}{2}\right)\ . \tag{4.29b}$$

The case $q_0 < 0$ corresponds to negative values of ρ and is disregarded. However, recent observations have sometimes been interpreted as favouring negative values of q_0. If confirmed, it would require either a non-vanishing Λ or some more drastic change in gravitation theory (Gunn and Oke 1975; Gunn and Tinsley 1975).

Combining equation (4.8) with (4.29) we get the desired relation

$$m_{\mathrm{bol}} = 5 \log \frac{1}{q_0^2} \{q_0 z + (q_0 - 1)[(1+2q_0 z)^{\frac{1}{2}} - 1]\} \qquad (4.30a)$$

$$-5 \log H_0 + M - 5 \quad (\text{for } q_0 > 0)$$

$$= 5 \log z\left(1 + \frac{z}{2}\right) - 5 \log H_0 + M - 5 \; (\text{for } q_0 = 0) \; . \qquad (4.30b)$$

This, of course, is written without considering the corrections due to evolution, heterochromatic absorption, and selective sensitivity of the plates.

Besides these there are also other factors which may affect the value of q_0. Thus it is sometimes held that the cosmic matter is primarily in the form of ionized hydrogen whose averaged-out density exceeds the density of luminous matter. If this is indeed so, then the light in course of its passage from one galaxy to another would suffer Thomson scattering. This leads to an attenuation of the light we receive from distant galaxies and thus would affect the m - z relation. Bahcall and May (1968) have calculated that this changes the observationally determined q_0 to $1{\cdot}15\, q_0$.

Dyer and Roeder (1972) consider the fact that, instead of a strictly uniform distribution of matter corresponding to the Robertson - Walker metric, we have concentrations as galaxies. This would affect the luminosity calculation and alter the value of q_0 (Roeder 1975). However, if the radii of the clumps be sufficiently small, the usual formula, rather than that given by Dyer and Roeder, would hold good (Weinberg 1976).

It has sometimes been held that the absolute magnitudes of the first few galaxies may depend on the number of galaxies in the cluster. There would then be a selection effect commonly called the Scott effect (Scott 1957) and this would introduce an error in q_0. However Sandage (1976) finds that the scatter in the absolute luminosity of the first few galaxies is extremely small ($\sigma \sim 0{\cdot}35$ mag).

In any case it seems difficult to accept any value of q_0 with confidence. There have been some attempts to set some bounds on q_0 (Rindler 1969), however these are so wide as to be of little use. (See also Landsberg and Brown, 1973.)

4.5. THE NUMBER COUNTS OR THE N - m RELATIONS

The optical observations, which are next in importance, are the studies of the number of galaxies above different luminosity levels. Here also, one can either adopt the series expansion procedure or use the field equations of general relativity.

The number of galaxies in the co-ordinate region 0 to r at any time can be written as

$$N = 4\pi n_0 \int_0^r \frac{r^2 \mathrm{d}r}{(1+kr^2/4)^3} \tag{4.31}$$

where n_0 is a constant. The constancy of n_0 both in space and time follows from the twin assumptions of uniformity and the co-moving nature of the co-ordinate system. Implicit also is the assumption that galaxies are neither being created nor destroyed at least within the space time region up to which our observation extends.

Again using a Taylor-series expansion, equation (4.31) gives

$$N = 4\pi n_0 \left(\frac{r^3}{3} - \frac{3}{20} kr^5 + \dots\right).$$

Substituting for r from equation (4.17), we get

$$N = \frac{4\pi n_0}{3} \frac{z^3}{H_0^3 R_0^3} \left[1 + \frac{3}{2} \frac{\alpha_2 - \alpha_1^2}{\alpha_1^2} z + \dots\right],$$

or,

$$\log_{10} N = \log \frac{4\pi n_0}{3R_0^3} + 3 \log \frac{z}{H_0} + \frac{3}{2} \times 0{\cdot}4343 \, \frac{\alpha_2 - \alpha_1^2}{\alpha_1^2} \, z + \ldots .$$

Eliminating z with the help of equation (4.19) we get the N - m relation as

$$\log_{10} N = \log \frac{4000\pi n_0}{3R_0^3} + 0{\cdot}6 \, (m-M) - 6{\cdot}2(2+K^*+\lambda^*)\alpha_1 \times 10^{0\cdot 2}(m-H)$$

where K^* is the part of K_0 linear in z. It is clear that from the ($\log N$-m) relation, that one can, in principle, determine the evolution factor λ^*.

However, the series-expansion method gives a coefficient involving α_2 and the curvature parameter k when one goes to the next term, i.e. the term involving $10^{0\cdot 4(m-M)}$ and one cannot determine both separately from number counts alone.

Using the field equations, Mattig (1958, 1959), gives the following formulae for the number of galaxies per square degree brighter than apparent magnitude m:

$$N = \frac{2\pi n}{QH_0^3} \begin{cases} (1-2q_0)^{-3/2}[P(1+P^2)^{\frac{1}{2}} - \sinh^{-1}P] & \text{for } k = -1 \\ (2q_0-1)^{-3/2}[\sin^{-1}P - P(1-P^2)^{\frac{1}{2}}] & \text{for } k = +1 \end{cases} ,$$

where n is the number of galaxies per unit volume, Q the number of square degrees in the sky, and

$$P = \frac{A[(2q_0-1)k]^{\frac{1}{2}}}{q_0(1+A) - (q_0-1)(1+2A)^{\frac{1}{2}}} \qquad \text{for } k = \pm 1 ,$$

$$A = H_0 \, 10^{0\cdot 2(m-M)}$$

and for $k = 0$,

$$N = \frac{4\pi n A^3}{3QH_0^3} \{\tfrac{1}{2}[1+A+(1+2A)^{\frac{1}{2}}]\}^{-3} .$$

For models with $q_0 \geq 1$ (k=1), N has a maximum. Sandage explains this maximum as similar to the maximum which one meets in the formula for the area of the spherical-cap out to radius

r:

$$S(r) = 2\pi R^2(1-\cos r/R)$$

at $r = \pi R$. For $r > \pi R$, the area decreases due to the vector r now approaching the origin after once describing the entire sphere. The maximum of N occurs at $dP/dA = 0$, i.e. at

$$A = \frac{(2q_0-1)+q_0(2q_0-1)^{\frac{1}{2}}}{(q_0-1)^2}\ .$$

It may be noted that, although the universe is closed also for $\frac{1}{2} < q < 1$, N has no maximum showing that, in these cases, one cannot extend one's observation to the antipoles. The expression for P may be written in the form

$$P = \frac{(2q_0-1)^{\frac{1}{2}}}{q_0(A^{-1}+1)+(1-q_0)(2A^{-1}+A^{-2})^{\frac{1}{2}}}$$

so that for $q_0 < 1$, P has a monotone increase tending to the value $(2q_0-1)^{\frac{1}{2}}/q_0$ as $A \to \infty$. T e existence of the limit shows that there is a horizon beyond which our observations cannot extend.

The N-m relations have been used by Hawkins and Martin (1977) to arrive at a value of the deceleration parameter $q_0 = 0{\cdot}8 \pm 0{\cdot}3$. (See however Brown and Tinsley, 1974; also Fennelly, 1976.)

4.6. THE ANGULAR MEASUREMENTS

We have already obtained in equation (4.3) an expression for ξ the distance by apparent size. Hence if we have a family of objects all of a standard proper size but at different distances then the observed angular size θ will be $\alpha\ \xi^{-1}$. Using equations (4.3) (4.4), and (4.8), we have

$$\theta = \text{const.}\ \frac{(1+z)^2}{D} = \text{const.} \times \frac{(1+z)^2}{10^{0\cdot 2(m-M)}}$$

with the help of equations (4.29) or (4.30) the above relation is translated into a relation between θ and z

$$\theta = \text{const.}(1+z)^2 H_0 q_0^2 \{q_0 z + (q_0 - 1)[(1+2q_0 z)^{\frac{1}{2}} - 1]\}^{-1} \quad (q_0 > 0)$$

$$= \text{const.}(1+z)^2 H_0 z^{-1}\left(1 + \frac{z}{2}\right)^{-1} . \quad (q_0 = 0)$$

It is easy to see that in both the cases θ considered as a function of z has a minimum. We have not considered the case of negative q_0 as this is not allowed by the general theory of relativity. However should the observations actually lead to a negative value of q_0, it is of interest to consider the corresponding relations. Of course one cannot then use the general relativity equations with vanishing Λ. A simple equation is obtained in case $q = -1$: (the steady-state case)

$$\theta = \text{const.}\ H_0(1+z)z^{-1} .$$

Obviously θ then has no minimum. A minimum is thus apparently a characteristic of relativistic models. (See Sachs, 1961; Liebes, 1964, Refsdal, 1964, 1966, Bertotti, 1966.)

However Sandage (1961a) points out that the angular diameter one obtains on a photographic plate may not show a minimum. The surface brightness of a galaxy does not remain constant as the galaxy recedes. One has for the surface brightness B

$$B = \frac{l}{\theta^2} = \frac{L\xi^2}{D^2} \times \text{const.} = \text{const.}(1+z)^{-4} .$$

Suppose now θ_i is the angle from the nucleus of a galaxy which is nearby and hence for it $z = 0$, at which the surface brightness is B_i. If the galaxy is now removed to a large distance the surface brightness will be reduced by the factor $(1+z)^4$. Consequently to find a point where the surface brightness is B_i, we must move close to the nucleus of the galaxy. Thus, as we go to galaxies at greater distances, the contours of equal surface-brightness move towards the nucleus.

Let us, for example, take the distribution of surface brightness across the face of a galaxy to be of the form

$$B = \frac{B_0}{(1+\theta/\alpha)^2} .$$

If $B_i \ll B_0$, we have approximately

$$\theta_i = \alpha \left(\frac{B_0}{B_i}\right)^{\frac{1}{2}} .$$

Taking B_i as constant, one ultimately gets θ_i = const. D^{-1} and this obviously shows no minimum. (For observational study of θ - z relations see Austin and Peach 1974.)

The influence of lumpiness in the distribution of matter has been investigated by Dyer and Roeder. They give the following formula for the distance by apparent size (Dyer and Roeder 1972; Gunn 1967)

$$H_0\xi = 3q_0^2(1-2q_0)^{-5/2}\left[\sinh^{-1}\left(\frac{1-2q_0}{2q_0}\right)^{\frac{1}{2}} - \sinh^{-1}\left\{\frac{1-2q_0}{2q_0(1+z)}\right\}^{\frac{1}{2}}\right]$$

$$- \frac{3q_0}{2}(1-2q_0)^{-2}\left[1 - \frac{(1+2q_0z)^{\frac{1}{2}}}{1+z}\right] + (1-2q_0)^{-1}\left[1 - \frac{(1+2q_0z)^{\frac{1}{2}}}{(1+z)^2}\right]$$

for $q_0 < 0{\cdot}5$

$$= \frac{2}{5}[1 - (1+z)^{-5/2}]$$

for $q_0 = 0{\cdot}5$

$$= 3q_0^2(2q_0-1)^{-5/2}\left\{\sin^{-1}\left(\frac{2q_0-1}{2q_0}\right)^{\frac{1}{2}} - \sin^{-1}\left[\frac{2q_0-1}{2q_0(1+z)}\right]^{\frac{1}{2}}\right\}$$

$$- \frac{3q_0}{2}(2q_0-1)^{-2}\left[1 - \frac{(1+2q_0z)^{\frac{1}{2}}}{1+z}\right] - \frac{1}{2}(2q_0-1)^{-1}\left[1 - \frac{(1+2q_0z)^{\frac{1}{2}}}{(1+z)^2}\right]$$

for $q_0 > 0{\cdot}5$.

In the above Λ has been taken to be zero (see also Zel'dovich, 1964). It will be noted that ξ has no maximum but tends to a finite limit as $z \to \infty$. Thus the apparent angular sizes would show a monotone decrease tending to a finite limit as $z \to \infty$. In a later paper Dyer and Roeder investigated the case where a fraction α of the matter density is in the form of intergalactic matter. Thus $\alpha = 0$ is the case presented above and $\alpha = 1$ corresponds to the usual homogeneous model. The authors obtained the expressions in closed form for the case $\alpha = 2/3$ as well. It was found that as α goes from unity to zero, the minimum angular size red-shift moves from 1·25 to infinity for $q_0 = \frac{1}{2}$. Thus the minimum appears to be due to the matter that lies in the path of the beam (Dyer and Roeder 1973; Zel'dovich 1964; Refsdal 1970; see also Weinberg 1976).

4.7. MORE GENERAL ANALYSIS OF THE OBSERVATIONAL DATA

All the analysis that we have so far discussed proceeds assuming the validity of the isotropic line element of Robertson and Walker with or without the field equations of general relativity. Kristian and Sachs (1966) have given an analysis which is free from this limitation. Essentially they assume, in general, only a Riemannian geometry for the space time and that light travels along null geodesics and obeys the area-intensity law. For some discussions they also use the field equations of general relativity taking the matter to be pressureless dust. The Kristian - Sachs analysis leads to the conclusion that the value of the deceleration parameter must be regarded with caution until better information on angular variation is available. Indeed constants of shear (i.e. anisotropy of expansion) and the incident electric type gravitational field may have considerable influence on q_0. (The incident electric type gravitational field is defined as the symmetric trace free-tensor of second rank obtained by contracting the Weyl tensor twice with the velocity vector, i.e. $E_{ac} = C_{abcd}\, u^b u^d$ where C_{abcd} is the Weyl tensor. In a way this is analogous to $\phi_{,\mu\nu} - \frac{1}{3}\delta_{\mu\nu}\nabla^2\phi$ in Newtonian theory where ϕ is the Newtonian gravitational potential.) Kristian and Sachs also gave some values for the upper bounds of shear and vorticity at our locale at the present epoch but even these bounds are subject to considerable uncertainty.

4.8. THE CURVATURE OF SPACE AND THE SEARCH FOR MISSING MATTER

Let us make a short summary of the situation. Equations (3.5) and (3.6) (p.22) may be rewritten in the following form for $p = 0$:

$$\frac{k}{R_0^2} = H_0^2\,(2q_0 - 1) \tag{4.32}$$

$$\frac{4\pi\rho_0}{3} = q_0 H_0^2 \,. \tag{4.33}$$

Equation (4.32) allows one to determine the curvature parameter if q_0 is known. Again in principle, q_0, H_0, as well as ρ_0 may be observationally determined and one can expect a check for equation (4.33). Unfortunately the value of H_0 suffers from an uncertainty about the distance scale (i.e. the absolute magnitude of the galaxies). Even then, of the three, the value of H_0 may be taken with the greatest confidence. As we have seen the position at present with q_0 is simply hopeless. A few years ago an estimate $q_0 \approx +1$ seemed fairly reliable - at least most cosmologists felt that they could assume that $q_0 > \frac{1}{2}$ so that $k = +1$ and the universe is closed. Today the opinions have swung to a value ~0 for q_0 and even negative values are not considered unlikely. But more than ever before, perhaps, we are conscious about the great uncertainty that hangs over the value of q_0. Again, the determination of ρ_0 depends on the doubtful assumption that matter in the universe is overwhelmingly in the form of visible galaxies. On this basis Oort (1970a,b,c) gave the value $\rho_0 = 3 \times 10^{-31}$ g cm^{-3} assuming $H_0 = 75$ km s^{-1} Mpc^{-1}. If instead one takes $H_0 = 50$ km s^{-1} Mpc^{-1}, the Oort value is reduced to $1{\cdot}4 \quad 10^{-31}$ g cm^{-3}. The Shapiro (1971) estimate is even lower: $\rho_0 \approx 5 \times 10^{-32}$ g cm^{-3}. However Einasto and colleagues (Einasto, Kaasik, and Saar 1974; Einasto, Saar, Kaasik, and Chernin 1974) have claimed that there is evidence of hidden matter concentrated about massive galaxies forming their coronas. The total mass of galaxies is then about an order of magnitude greater than the mass of their visible parts and this would correspondingly raise the density values of Oort and Shapiro. From different considerations Ostriker, Peebles, and Yahil (1974) have also arrived at similar values for the density. However the validity of the ideas of Einasto *et al.* as well as those of Ostriker *et al.* has been questioned by Burbidge (1975).

The Oort-Shapiro values of the density would lead to an open universe with a small q_0 in agreement with present estimates. (Even with a small positive or negative q_0 one can have a closed universe and high values of ρ_0 if one admits the cosmological term as has indeed been suggested by Gunn and Tinsley 1975.) However there was (and to some extent it still persists) a bias in favour of closed universes amongst a school of cosmologists. For this to be true one must assume that the matter density of the universe exists predominantly in non-luminous form. Presumably this may be in the form of neutrinos, gravitational waves, (the gravitational waves do not contribute directly to the energy-momentum tensor, nevertheless they may effectively increase the mass density by a viscosity effect as studied by Hawking 1966), black holes (Lynden-Bell 1969; Wolfe and Burbidge 1970), intergalactic dust (see however the negative results of the calculations by Nickerson and Partridge (1971) and Crane and Hoffman (1973)), or gas.

Before examining these possibilities, we would note some semi-philosophical arguments that have been put forward in this connection. As we have already seen, according to some the very elusive Mach principle requires a closed universe. Again Collins and Hawking (1973a) gave an intriguing argument in favour of $k = 0$. The high degree of isotropy of the microwave background indicates a remarkable isotropy and homogeneity of the universe. But we have inhomogeneities as well in the form of galaxies. Now one may expect that the galaxies have been formed by the growth of perturbations in an isotropic homogeneous background. Collins and Hawking suggest, but do not exactly prove, that such a formation of galaxies can take place only if the universe is expanding just fast enough to avoid recollapse, i.e. $k = 0$.

Let us return now to the question of missing matter that can close the universe or make $k = 0$. Cowsik and McClelland (1973) have claimed that if the neutrino rest-mass be of the order of a few electron volts, then they may close the universe (see also Bludman 1976). A non-zero rest-mass of neutrinos has sometimes been suggested as a possible explanation of the failure of Davis, Harmer, and Hoffmann (1968) to detect solar neutrinos of expected intensity. Regarding black holes, Press and Gunn (1973) concluded from the absence of predicted gravitational focusing effects that the black holes of galactic mass cannot provide sufficient mass for closure. Again, considering the evaporation of black holes by the Hawking mechanism, Carr (1975) finds that the observed gamma radiation intensity indicates that primordial black holes of mass around 10^{15}g are quite inadequate for closure of the universe. (See Thorstensen and Partridge, 1975.) There thus remains only the possibility of inter- and intragalactic hydrogen of sufficient density. This is at once the most conservative and the most promising possibility.

If hydrogen is in the atomic form, one would expect to observe absorption features in the spectra of quasi-stellar objects assuming that they are at distances corresponding to a cosmological interpretation of their red shifts. A photon of frequency ν_e emitted at time t_1 is received at time t_0 as a photon of degraded frequency ν_0 where $\nu_0 = \nu_e/(1+z) = \nu_e\, R(t_1)/R(t_0)$. If now in the intervening region there is hydrogen with a characteristic absorption frequency ν_a and if $\nu_e \geq \nu_a \geq \nu_0$, then there will be a strong absorption in the region of the path where the photon frequency has decreased to ν_a. Along with the absorption there will also be spontaneous and stimulated emission at the same frequency. However the spontaneous emission, being isotropic, would only increase the general background intensity and so need not be considered for the directed beam. Thus the change of photon number along the path of the beam is given by

$$\frac{dn}{dl} = -\sigma n n_H \,(1 - e^{-h\nu/KT})$$

where we have used the Einstein relation between the absorption and stimulated emission; dl is the element of proper path length, σ is the frequency-dependent absorption cross-section, n and n_H are respectively the number density of photons and hydrogen atoms in the particular region.

Replacing dl by $c dt$ (for the moment we are not using

the special choice of units to have $c = 1$) we can formally integrate the above equation:

$$n = n_0 \exp[-\int_0^n \sigma n_{\rm H}(1-e^{-h\nu/KT})\,dt]$$

$$= n_0 \exp[+\int_{\nu_e}^{\nu_0} c\sigma n_{\rm H}(1-e^{-h\nu/KT})\,d\nu/\left(\frac{d\nu}{dt}\right)].$$

The resonance absorption cross-section is of the form

$$\sigma = \frac{\pi e^2}{mc} fg(\nu-\nu_{\rm a})$$

where f is the appropriate oscillator strength and $g(\nu-\nu_{\rm a})$ is the profile function which may be replaced by the Dirac δ-function without sensible error. We thus obtain for the optical depth τ (this is defined such that $\exp(-\tau)$ is the attenuation factor for photon intensity)

$$\tau = \frac{\pi e^2 f}{m\nu_{\rm a}} (n_{\rm H})_a \,(1-e^{-h\nu_{\rm a}/KT_{\rm a}})/\left(\frac{\dot{R}}{R}\right)_{\rm a}$$

where we have used the relation νR = const. The subscript 'a' stands for values at the locale where the photon in transit has the frequency $\nu_{\rm a}$. Using now equation (3.31) of (p.43) for $(\dot{R}/R)_{\rm a} \equiv H_{\rm a}$, we get

$$\tau = \frac{\pi e^2 f}{m\nu_0 H_0} (n_{\rm H})_{\rm a} \,(1-e^{-h\nu_{\rm a}/KT_{\rm a}})(1+z_{\rm a})^{-1}(1+2q_0 z_{\rm a})^{-\frac{1}{2}}$$

where $z_{\rm a} = (\nu_{\rm a}/\nu_0 - 1)$.

The above expression for τ is valid only if somewhere in its path the frequency of the photon had been $\nu_{\rm a}$, i.e. if $\nu_0 \le \nu_{\rm a} \le \nu_{\rm e}$ or in terms of z, $z \ge z_{\rm a} \ge 0$. Outside this region there is no absorption and stimulated emission, and consequently no attenuation, i.e. τ vanishes. One thus expects an absorption trough in the region ν_0 given by $\nu_{\rm a}(1+z)^{-1} \le \nu_0 \le \nu_{\rm a}$.

Attempts have been made to determine the concentration of atomic hydrogen by using the expression for τ in case of the Lyman α line and the 21 cm microwave line. The Lyman α line arises from the transition $2p \to 1s$, the oscillator strength $f = 0{\cdot}416$, and the frequency $\nu_a = 2{\cdot}46 \times 10^{15} s^{-1}$. The value of $h\nu_a/k$ comes out as $\sim 10^5$ K, so that the exponential term corresponding to stimulated emission is negligible if $T_a \ll 10^5$. (If on the other hand T_a be comparable to 10^5 K, the degree of ionization of hydrogen would be very high and Lyman α absorption would be negligible.) Neglecting the exponential term and remembering $H_0 = 50$ km s^{-1} Mpc^{-1}, we get finally for τ

$$\tau = \frac{8\times10^{10}(n_H)_a}{(1+z_a)(1+2q_0z_a)^{\frac{1}{2}}} .$$

For the quasar 3C9, (z 2·01), Gunn and Peterson (1965) believe to have observed a positive effect with $\tau = 0{\cdot}5$ for the absorption in the neighbourhood of the quasar (i.e. at $z = 2$) but later observers (see Field 1969; Burbidge and Burbidge 1967) have not found any positive effect either for 3C9 or for any other quasar. Subsequently the Gunn-Peterson value has been interpreted as an upper limit. As we are interested in the question of closure we must have $q_0 > \frac{1}{2}$. Taking $q_0 = \frac{1}{2}, z = 2, \tau < 0{\cdot}5$, we get $n_H < 3\times10^{-11}$ cm^{-3} at $z = 2$. The closure would however require $n_H \approx 10^{-5}$ cm^{-3}.

The 21-cm line arises from a spin-flip transition in the 1s state from total spin zero to total spin unity, i.e. from a state in which the spins of the proton and the electron are antiparallel to one in which they are parallel. The value of $h\nu_a/k$ for this line is 0·068 K and hence the exponential term may be replaced by $h\nu_a/kT_s$ where T_s is the 'spin excitation temperature'. Thus in this case with appropriate values of f, ν_a and H_0

$$\tau = 3 \times 10^4 \, n_H \, T_s^{-1} .$$

Unfortunately the observational results by different investigators (Field 1959; Koehler and Robinson 1966; Penzias and Scott 1968) are not concordant. Penzias and Scott gave $\tau < 5 \times 10^{-4}$, so that

$$\frac{n_H}{T_s} < 1{\cdot}6 \times 10^{-8} \text{ cm}^{-3} \text{ deg}^{-1} .$$

If one takes $T_S \approx 2{\cdot}7$ K (the background radiation temperature) then $n_H < 5 \times 10^{-8}$ cm^{-3} and this again falls far short of the value required for closure. However T_S is rather uncertain. If there was a reheating of the gas leading to a significant ionization of hydrogen, then the Lyman α line arising from recombination may heat up the gas to high temperatures. Indeed if $T_S \gtrsim 600$ K, then the absence of 21-cm absorption may not be inconsistent with the requisite density of hydrogen.

However with T_S sufficiently high, one would expect a 21-cm emission. The microwave background-radiation, as it passes through the hydrogen would then suffer change at this wavelength - the net effect being an enrichment of the radiation at this wavelength. However, owing to red shift, the enrichment would be spread over all longer wavelengths and ultimately one would expect a 'step' in the background-radiation intensity at the corresponding frequency. The magnitude of this step would obviously be related to the hydrogen density - it would be given by

$$\Delta N = N_{\nu_a - 0} - N_{\nu_a + 0} = 8\pi\nu_a H_0^{-1} c^{-2} n_H(t_0) \frac{\pi e^2}{mc} f$$

where ΔN is the jump in photon number-density per unit frequency-interval and $n_H(t_0)$ is the number density of hydrogen atoms at the epoch of observation. Attempts to observe such a step have not been successful and one can thereby set an upper limit to the possible value of $n_H(t_0)$ (Penzias and Wilson 1969)

$$n_H(t_0) < 2 \times 10^{-6} \text{ cm}^{-3} .$$

(This is not exactly the limit given by Penzias and Wilson; this is the value when we use $H = 50$ km s^{-1} Mpc^{-1}.) Although this limit is also below the value required for closure, as it is rather close to that value, it is usually concluded that the possibility of hot hydrogen of sufficiently high density is not completely ruled out.

After this consideration of atomic and ionized hydrogen we may take up the question of molecular hydrogen. Attempts have been made to detect absorption features due to molecular hydrogen in the spectrum of 3C9. This leads to an upper limit (Field, Solomon, and Wampler 1966):

$$n_{H_2} < 4\cdot 10^{-9}\ \mathrm{cm}^{-3}\ .$$

Again it is several orders of magnitude less than the critical value (see also Noonan 1972).

Many have found it a little difficult to believe that the amount of intergalactic hydrogen is really so small as given by the absence of absorption features in the spectra of 3C9 and other quasars of large red-shift. Instead they have argued that this is only an evidence against the hypothesis of the cosmological origin of quasar red-shifts. However this is not the place to enter into this complicated and controversial question, we would only remark that it seems difficult to believe in the presence of hydrogen in sufficient density in the absence of all positive corroboration.

The question of ionized hydrogen may be pursued a little further. The electrons set free by the ionization of hydrogen would interact with radiation primarily by Thomson scattering. This leads to an attenuation of any directed beam, the optical depth up to a co-ordinate r being

$$\tau = \int_0^r \sigma_T\, n_e \mathrm{d}l$$

where $\mathrm{d}l$ is the element of proper path length ($\neq \mathrm{d}r$), σ_T is the Thomson scattering cross section for electrons (and is independent of the frequency) and n_e is the number density of electrons, assumed to be the same as that of protons.

Changing the variable from the path length l to z,

$$\mathrm{d}l = c\mathrm{d}t = \frac{c\mathrm{d}R}{HR} = \frac{c\mathrm{d}z}{H_0(1+z)^2}\ (1+2q_0 z)^{-\frac{1}{2}}\ .$$

Also,

$$n_e = n_{H+} = \frac{\rho}{m_H} = \frac{3q_0 H_0^2}{4\pi G m_H}\ (1+z)^3$$

where we are retaining c and G. The optical depth thus

becomes on integration

$$\tau = \frac{H_0 c \sigma_T}{4\pi G m_H} \frac{1}{q_0} [(3q_0 + q_0 z - 1)(1 + 2q_0 z)^{\frac{1}{2}} + 1 - 3q_0] . \qquad (4.34)$$

The optical depth for a given H_0 thus depends on q_0. Again remembering that our interest is in closure of the universe which would mean $q_0 \geq \frac{1}{2}$, we note that with H_0 = 50 km s^{-1} Mpc^{-1} and $q_0 = \frac{1}{2}$, the optical depth is unity at z = 3·5 (Sciama (1971) gave z = 7 as he used H = 100 km s^{-1} Mpc^{-1}.) Thus, with our values, the attenuation would not be significant even up to z = 2.

One may expect an isotropic X-ray background from the hot plasma due to thermal bremsstrahlung. The characteristic of a thermal bremsstrahlung-spectrum is the exponential decrease of emissivity with photon energy. Thus the emissivity for free-free transition is

$$Cz^2 N_e N_i g T^{-\frac{1}{2}} e^{-h\nu/KT}$$

where $C = 5{\cdot}44 \times 10^{-39}$ cgs units, g is the Gaunt factor, may be taken to be nearly unity, z is the charge number of nucleus involved, N_e is the number density of electrons and may be taken equal to 1·2 N_H (considering the intergalactic gas to contain 10 per cent helium), and N_H is the number density of protons. Hence

$$\Sigma z^2 N_i N_e g \approx 1{\cdot}85\, N_H^2 .$$

Now, as the medium within our galaxy is opaque to soft X-rays with photon energy less than 0·1 keV and as the exponential factor would cause a rapid decrease for photon energies above kT, we can expect to observe the X-rays due to thermal bremsstahlung only if the temperature of the gas be such that $kT \geq 0{\cdot}1$ keV or $T > 10^6$ K. Thus should observations reveal an isotropic X-ray background obeying an exponential law, one would be able to calculate the density and temperature of the gas from a study of the energy distribution.

However it is often difficult to decide between an exponential and a power-law spectrum and there has been lengthy controversies between the protagonists of thermal bremsstrahlung and inverse Compton process theories as to the origin of the observed X-ray background. Observations have revealed a diffuse X-ray background extending from ~0·25 keV to 100 MeV. The high degree of isotropy of this background indicates an extragalactic origin. The general characteristic of the X-ray spectrum may be described in the following manner. For energies lying between 1 to 40 keV, the intensity follows a -0·8 power-law, from 40 keV to 1 MeV it is a -1·3 power-law. Below 1 keV in the range ~250 eV there seems to be a sharp increase in flux and also from 1 to 5 MeV the intensity is significantly higher than that given by the -1·3 power-law (Silk 1970). It is still not quite clear whether these observations are consistent with a simple inverse Compton process between the low-energy photons of the microwave background and high-energy electrons. In the earlier calculations an apparently striking agreement was obtained with the observational data, the peculiarities of the observed flux being correlated with the peculiarities believed to exist in the electron spectrum (Felton and Morrison 1966; Brecher and Morrison 1969). Cowsik and Kobetich (1972) reinvestigated the inverse Compton X-ray spectrum taking into account (i) the isotropy of the background radiation, (ii) the Planck distribution of the energy of photons, and (iii) the differential scattering cross-section for the Compton process. They arrived at the conclusion that the Compton process is inadequate to account for the observed X-ray flux at low energies, especially the break at 40 keV. The difference in the energy range 20 - 200 keV between the observed and estimated Compton intensity could be explained as due to thermal bremsstrahlung from an intergalactic distribution of hydrogen at temperature $\sim 3 \times 10^{8}$ K and having a number density $\sim 3 \times 10^{-6}$ cm^{-3}. Silk and Tarter (1973) reached the conclusion that a substantial fraction of the observed X-ray background may be due to cluster of galaxies. If clusters are bound by intracluster gas they may be X-ray sources and may have gas enough to close the universe. However the considerations of Silk and Tartar are highly tentative.

However, the peculiarities of the X-ray spectra, which,

according to Cowsik and Kobetich, cannot be reconciled with a simple Compton process, have recently become questionable. Thus there are indications that the excess at ~260 keV may be of galactic origin (McCammon, Bunner, Coleman, and Kraushaar 1971; Strittmatter, Brecher, and Burbidge 1972) and the Apollo 15 observations have considerably reduced the excess in the region 1 - 5 MeV (Trombka, Metzger, Arnold, Matheson, Reedy, and Peterson 1973). Even the sharp change in spectral index at 40keV has been questioned (Kasturirangan and Rao 1972; Dennis, Suri, and Frost 1973). Thus it may well be that the entire range 20 keV to 1 MeV may be represented by a single power-law although a flattening above 1 MeV seems to be confirmed (Kuo, Frye, and Zych 1973; Hopper, Mace, Thomas, Abbots, Frye, Thomson, and Staib 1973) and there may be a contribution from red-shifted photons from neural pion decay (Vette, Gruber, Matheson, and Peterson 1970; Stecker 1973a,b).

Brecher (1973) on the other hand has argued that the peculiarities, even if they exist, are not irreconcilable with a Compton model. He points out that there is a basical identity between the inverse Compton process and the synchroton mechanism so that electrons with a given spectral distribution produce parallel Compton X-ray and synchroton radio-spectra. As there is a close parallelism between the diffuse X-ray spectrum we observe and the radio spectra of 3C 220·3 and 3C 255 (Brecher 1973; Kellermann, Pauliny-Toth, and William 1969) it was argued that the observed X-ray-spectra can conceivably be generated by a suitably bent electron-spectra.

To sum up, the situation remains confused and the extreme high temperature assigned to the intergalactic hydrogen in the Cowsik - Kobetich calculation seems difficult to reconcile with the plausible reheating processes and the lack of sensible departure of the background microwave-radiation from Planck distribution. Further, as the bremsstrahlung radiation is proportional to ρ^2, so any clumpiness in the gas distribution would lead to much lower densities (Silk 1973a).

5
RELATIVISTIC MODELS NOT OBEYING THE COSMOLOGICAL PRINCIPLE

5.1. THE MOTIVATION BEHIND STUDY OF THESE MODELS

There are fairly strong grounds which lead one to study the cosmological solutions of the field equations of general relativity even though they may not conform to the cosmological principle. Firstly the isotropic models form a 'set of measure zero' amongst the totality of all models permitted by the equations of general relativity. Again the uniformity that is postulated in the cosmological principle is most certainly lacking on a small scale and many entertain doubts as to its validity even on a large smoothed-out scale. Besides, as we have already pointed out, the existence of particle horizons in the isotropic universes makes it difficult to understand how a uniformity may obtain between regions at epochs when there has been no possibility of previous communication.

However, once we give up the powerful cosmological principle, not only do we no longer have a simple and rigid mathematical frame work (i.e. the Robertson-Walker metric) in which to pursue our investigations but we are simply lost in the midst of the diversity and complexity that emerges. In this background two courses seem to be open - one may try to deduce general formulae and theorems without bothering about a complete solution of the field equations representing a fully fledged cosmological model, or else one may again adopt some principle, which, although weaker than the cosmological principle, nevertheless allows one to build up concrete models. To the first class belong the results obtained by Raychaudhuri, Ehlers, and others - the most noteworthy being the singularity theorems of Hawking and Penrose. In the second group we have a very large number of investigations on homogeneous but anisotropic models and spherically-symmetric or cylindrically-symmetric solutions.

5.2. SOME GENERAL FORMULAE AND THEOREMS

The Riemann-Christoffel tensor may be defined by the relation:

$$2v^{\mu}{}_{;[\alpha;\beta]} = R^{\mu}{}_{\nu\beta\alpha}\, v^{\nu} \tag{5.1}$$

where v^{μ} is any arbitrary vector, the semicolon denotes covariant derivative and the square brackets indicate antisymmetrization. In what follows we shall often consider the vector v^{μ} to be the velocity vector of a continuous matter distribution in hydrodynamic motion. (Many of the formulae however are valid for any arbitrary choice of v^{μ}.) v^{μ} is thus a unit time-like vector. The tensor $v_{\alpha;\beta}$ may be split up into the following physically significant quantities:

(i) The scalar of expansion $\theta \equiv v^{\mu}{}_{;\mu}$. It gives the rate of dilation of a three-space element locally orthogonal to the vector v^{μ}.

(ii) The acceleration $\dot{v}_{\alpha} \equiv v_{\alpha;\beta}v^{\beta}$. This gives the departure of the velocity field from geodesicity. In the absence of non-gravitational interactions $\dot{v}_{\alpha}$ would vanish.

(iii) The shear tensor

$$\sigma_{\alpha\beta} \equiv v_{(\alpha;\beta)} - \tfrac{1}{3}(g_{\alpha\beta} - v_{\alpha}v_{\beta})\theta - \dot{v}_{(\alpha}v_{\beta)}\,. \tag{5.2}$$

This is a symmetric tensor, is trace free, orthogonal to the vector v^{μ}, and gives the change of shape of space elements orthogonal to v^{μ}.

(iv) The vorticity tensor

$$\omega_{\alpha\beta} \equiv v_{[\alpha;\beta]} - \dot{v}_{[\alpha}v_{\beta]} \tag{5.3}$$

This is an antisymmetric tensor and vanishes if the velocity vector is hypersurface orthogonal. One can also define the vorticity vector

$$\omega^{\mu} \equiv \tfrac{1}{2}\eta^{\mu\nu\rho\sigma} v_{\nu} v_{\rho;\sigma} .$$

The vorticity vector (or equivalently the tensor $\omega_{\mu\nu}$) is a measure of the rotation of the local rest-frame relative to the compass of inertia (Synge 1937; Gödel 1949).

If now one contracts equation (5.1) by putting $\mu = \beta$ and then further contracts with v^{α}, one obtains a scalar equation:

$$2v^{\mu}{}_{;[\alpha;\mu]} v^{\alpha} = R_{\nu\alpha} v^{\nu} v^{\alpha}. \tag{5.4}$$

Using the definitions of expansion, acceleration, shear, and vorticity that we have just given, we transform the l.h.s. to get

$$\theta_{,\alpha} v^{\alpha} + \frac{1}{3}\theta^{2} - \dot{v}^{\alpha}{}_{;\alpha} + 2(\sigma^{2}-\omega^{2}) = -R_{\mu\nu} v^{\mu} v^{\nu}. \tag{5.5}$$

In obtaining (5.5) we have used the following relations

$$\dot{v}_{\mu} v^{\mu} = 0$$

$$\sigma_{\mu\nu} v^{\nu} = \omega_{\mu\nu} v^{\nu} = 0$$

$$\sigma^{2} \equiv \tfrac{1}{2}\sigma_{\mu\nu}\sigma^{\mu\nu}$$

$$\omega^{2} \equiv \tfrac{1}{2}\omega_{\mu\nu}\omega^{\mu\nu}$$

The relation (5.5) is an identity for any unit vector v^{μ}. We now use the field equations and the expression for the energy-stress tensor (equations 3.1 and 3.4 of (p.22)) in the r.h.s. of (2.5) to obtain

$$\theta_{,\alpha} v^{\alpha} + \frac{1}{3}\theta^{2} - \dot{v}^{\alpha}{}_{;\alpha} + 2(\sigma^{2}-\omega^{2}) + 4\pi(\rho+3p) = 0 . \tag{5.6}$$

Both (5.5) and (5.6) are usually referred to as the Raychaudhuri equation. These are scalar equations. A vector equation of some interest may be obtained from (5.1) if one simply contracts by putting $\mu = \beta$.

$$\frac{2}{3}\,\theta_{,\alpha}\,(g^{\alpha\beta}-v^{\alpha}v^{\beta}) = \sigma^{\alpha\beta}_{;\alpha} + \sigma^{\alpha\beta}\dot{v}_{\alpha} + 2\sigma^{2}v^{\beta}$$

$$+ \eta^{\mu\nu\alpha\beta}\,(\omega_{\mu,\alpha}\,v_{\nu}-2\omega_{\mu}v_{\nu}p_{,\alpha}/p+\rho) \qquad (5.7)$$

where, as before $\eta^{\mu\nu\alpha\beta}$ is the completely antisymmetric tensor. Several interesting theorems follow very easily from the above equations (e.g. Raychaudhuri 1955a; Ehlers 1961; Banerji 1968). These are stated without details of the proofs.

(i) If the motion be irrotational ($\omega_{\mu}=0$) and geodetic ($\dot{v}_{\alpha}=0$) (which will be the case if the space components of pressure-gradient vector vanishes), the expansion scalar will become arbitrarily large in a finite time, either in the past or in the future. In this state the density ρ will also be infinite.

(ii) Of all the expanding, rotation free, dust cosmological models specified by a given value of ρ and θ at a particular epoch, the time scale to the singular state of infinite density will be a maximum for the isotropically expanding models (i.e. for models with $\sigma_{\mu\nu} = 0$).

From this theorem, it may appear that so far as the time span is concerned, nothing will be gained by changing over to anisotropic models. However, in practice, it is difficult to make an accurate determination of ρ. As we have seen in the last chapter it is more promising to attempt a determination of the deceleration parameter which is linked up with $\theta_{,\alpha}v^{\alpha}/\theta^{2}$. In case θ and the deceleration parameter are specified, one can build up models with somewhat greater time span by introducing anisotropy as has been shown, for example, by Kantowski and Sachs (1966).

(iii) For a distribution of perfect fluid, if the density be spatially uniform with the motion irrotational, the expansion will also be spatially uniform. Further, if there is spherical symmetry, the shear will vanish everywhere. This theorem is due to Misra and Srivastava (1973). We give below the sketch of a proof.

The uniformity of density means

$$\rho_{,\mu} = \alpha v_{\mu} \tag{5.8}$$

so that

$$\rho_{,\mu}(g^{\mu\nu} - v^{\mu}v^{\nu}) = 0 \tag{5.9}$$

$$\rho_{,\mu}\, v^{\mu}_{;\nu} = 0 \ . \tag{5.10}$$

Differentiating (5.9) covariantly along the world line we get using equation (5.10)

$$\rho_{,\mu;\beta}\, v^{\beta}(g^{\mu\nu} - v^{\mu}v^{\nu}) = \rho_{,\mu} v^{\mu} \dot{v}^{\nu} \ . \tag{5.11}$$

The divergence relation $T^{\mu\nu}_{;\mu} = 0$ gives

$$\theta = \rho_{,\mu} v^{\mu} \frac{1}{(p+\rho)} \tag{5.12}$$

$$\dot{v}^{\mu} = -\frac{p_{,\nu}}{(p+\rho)}\,(g^{\mu\nu} - v^{\mu}v^{\nu}) \ . \tag{5.13}$$

Differentiating equation (5.12) and using (5.10), (5.11), and (5.13) we get

$$\theta_{,\alpha}(g^{\alpha\beta} - v^{\alpha}v^{\beta}) = 0 \ , \tag{5.14}$$

i.e. the expansion is spatially uniform. For the second part of the theorem we note that in spherically symmetric spaces, the shear tensor has only one independent component. For, if one writes the line element in the form

$$ds^2 = e^{\nu}dt^2 - e^{\lambda}dr^2 - e^{\omega}(d\theta^2 + \sin^2\theta d\theta^2)$$

where λ, ω, and ν are functions of r and t, only the three diagonal components of the shear tensor survive. But the trace is zero and because of symmetry $\sigma_{\theta\theta} = \sigma_{\phi\phi}$. Hence there is only one independent component. Looking back at equation (5.7) and using (5.14) we get

$$\sigma^{\alpha\beta}_{\ ;\alpha} + \sigma^{\alpha\beta}\dot{v}_{\alpha} + 2\sigma^{2} v^{\beta} = 0 \ . \qquad (5.15)$$

As there is only one independent component of $\sigma_{\mu\nu}$ and the derivative with respect to r alone exists, one has from equation (5.15) that $\sigma_{\mu\nu}$ will vanish everywhere if it vanishes for some value of r. The condition of regularity of the field at the centre of symmetry demands the vanishing of shear there - hence the shear vanishes everywhere.

(iv) If the motion be shear free, geodetic, and irrotational, and the energy stress tensor be that for a perfect fluid, the universe is completely isotropic and spatially homogeneous. A somewhat analogous but, in a way, stronger theorem is due to Liang (1975) who instead of geodesicity of the world lines assume a functional relation between the pressure and density and arrives at the same result.

(v) The only dust universes exhibiting rigid rotation and spatial homogeneity are the Einstein static universe and the Gödel stationary model. We shall consider the Gödel solution in a later chapter. The theorem was stated by Gödel and a detailed proof, which is quite complicated and depends on the existence of a transitive group of motions, is due to Ozsvath (1965). We note, however, that both these models require a non-vanishing cosmological term.

(vi) Spatially homogeneous universes having non-vanishing rotation and expansion must have a non-vanishing shear. This theorm is true for dust and perfect-fluid universes (Schucking 1957; Banerji 1968; King and Ellis 1973).

Several more formulae of wide generality may be written out. A list of these along with the corresponding formulae in Newtonian cosmology is given by Ellis (1971). We note here two comparatively simple relations for dust distributions (Raychaudhuri 1955b).

$$8\pi\rho = -\frac{R^*-2\sigma^2}{2} + \frac{\theta^2}{3} \tag{5.16}$$

$$(R^* + 2\sigma^2)_{,\alpha}v^\alpha = -(12\sigma^2 - 2R^* - 8\omega^2)\,\frac{\theta}{3} \tag{5.17}$$

where R^* is the Ricci scalar for the three-space locally orthogonal to the world lines.

5.3. HOMOGENEOUS UNIVERSES

The theory of continuous groups of motions leads to an elegant simplification in the study of cosmological models having homogeneous space sections. If there is a three-parameter group which is simply transitive on the homogeneous varieties, then Taub (1951) and Heckmann and Schucking (1962) have shown how the field equations of Einstein can be reduced to ordinary differential equations with time as the independent variable. The time lines are here defined as the orthogonals to the homogeneous varieties and are geodetic. They may not be the world lines of matter - indeed if vorticity is present, the velocity vector is not hypersurface orthogonal and hence must be different from the time-like vector which is orthogonal to the homogeneous varieties.

The differential equations, as presented by Taub or Heckmann and Schucking involve the structure constants of the group. However there can be only nine non-equivalent sets of structure constants for these groups, called the Bianchi groups (Estabrook, Wahlquist, and Behr 1968; Ellis and MacCallum 1969). One can thus attempt a systematic investigation of all the different spatially homogeneous models. Ozsvath has further shown that in a number of cases these field equations for the homogeneous models can be derived from a variational principle and one can thus use to advantage the methods of Hamiltonian mechanics.

For the case where the universe, although spatially homogeneous, nevertheless does not permit a simply transitive three-parameter group of motions, Kantowski and Sachs (1966) have given some solutions (see also Kohler 1933). The solutions are characterized by an absence of rotation, but there is a non-vanishing shear which becomes dominating at early stages of the universe (i.e. near the singularity).

Returning to the case where there is a simply transitive three-parameter group of motions, we write down the field equations as given by Heckmann and Schucking

$$\tfrac{1}{2}[k^a_b k^b_a - (k^a_a)^2 - R^*] = -8\pi\rho u^2 \tag{5.18}$$

$$k^a_b c^b_{ac} - k^a_c c^b_{ba} = -8\pi\rho u u_c \tag{5.19}$$

$$(k^b_a)^\cdot + k^c_c\, k^b_a - [(k^c_c)^\cdot + \tfrac{1}{2} k^c_d\, k^d_c + \tfrac{1}{2}(k^c_c)^2]\delta^b_a + R^{*b}_{\ a} - \tfrac{1}{2}R^*\delta^b_a$$
$$= -8\pi\rho u_a u^b\ . \tag{5.20}$$

The line element is of the form

$$ds^2 = dt^2 + g_{ik}\, dx^i\, dx^k\ .$$

The roman indices run from 1 to 3 and we have the following defining relations

$$k^a_b \equiv \tfrac{1}{2}\,\gamma^{ac}\,\dot{\gamma}_{bc}$$

$$\gamma_{ab} \equiv g_{ij}\xi^i_a\xi^j_b\ .$$

$$\gamma_{ab}\gamma^{bc} \equiv \delta^c_a$$

$$u^2 \equiv 1 - u^a u_a$$

$$u_a \equiv u_j\xi^j_a$$

$$\chi_a\chi_b - \chi_b\chi_a \equiv C^c_{ab}\,\chi_c$$

$$\chi_a \equiv \eta^i_a \frac{\partial}{\partial x^i} \quad .$$

The vectors η^i_a are the generators of the group, $C^a_{b\,c}$ are called the constants of structure of the group; they are antisymmetric in the lower pair of indices and satisfy the Jacobi identity

$$C^e_{ab}\,C^f_{ec} + C^e_{bc}C^f_{ea} + C^e_{ca}C^f_{eb} = 0 \;.$$

R^*_{ij} is the Riemann-Christoffel tensor for the three-space metric g_{ij} and

$$R^*_{ab} = R^*_{ij}\;\xi^i_a\xi^j_b$$

$$R^* = \gamma^{ab}R^*_{ab} \;.$$

All the terms in equations (5.18) - (5.20) including ρ are functions of t alone. The various possible sets of C^a_{bc} for the nine different Bianchi groups are listed in the Appendix.

Ozsvath (1970, 1971) has shown that (5.20) may be expressed as a Euler-Lagrange variational principle if the constants of structure satisfy the condition $C^a_{bc} = 0$ if either $a = b$ or $a = c$. To this class belongs the Bianchi groups of types I, II, VII, and IX as well as two special cases of types VI and VII. The Lagrangian is given by

$$L = \gamma^{\frac{1}{2}}[k^a_bk^b_a - (k^c_c)^2 + R^*] - 16\pi\rho\gamma^{\frac{1}{2}}u^2$$

and the corresponding Hamiltonian is

$$H = \gamma^{\frac{1}{2}}[k^a_bk^b_a - (k^c_c)^2 - R^*] + 16\pi\rho\gamma^{\frac{1}{2}}u^2 \;.$$

As the Lagrangian does not explicitly involve the time, the Hamiltonian has a constant value, which in view of (5.18) is zero. The Ozsvath programme is to compute the Hamiltonian and write down Hamilton's canonical equations. He then introduces a suitable transformation of the time variable such that the field equations, which are now ordinary first-order differential equations with respect to the new variable, are analytic. 'The problem is then solved in the sense that everything else can be done by computers and by the application of the qualitative theory of differential equations.' For a detailed discussion and examples of how the programme works, the original papers by Ozsvath may be consulted.

The two types of spatially homogeneous models that have received the greatest attention are those belonging to the Bianchi type I and type IX. The Bianchi type I is abelian (the group of three independent translations) and allows for an expansion and shear but does not permit any vorticity. Mathematically this is the simplest type and the field equations can be integrated with little difficulty. For the empty universe belonging to this type one has the well known Kasner solution (Kasner, 1927).

$$ds^2 = dt^2 - t^{2p_1}dx^2 - t^{2p_2}\,dy^2 - t^{2p_3}dz^2$$

with

$$p_1+p_2+p_3 = p_1^2+p_2^2+p_3^2 = 1 \ .$$

Except for the triplet (001), one of the three ps is negative and the other two positive. In the first case the space time is flat and in all other cases there is a physical singularity as $t\to 0$ and $t\to\infty$ (Joseph 1957). Biachi type I universes containing matter and/or electromagnetic fields have been investigated by a fairly large number of workers (e.g. Raychaudhuri 1958; Heckmann and Schucking 1962; Kompaniets and Chernov 1969; Thorne 1967; Robinson 1961; Saunders 1969, Jacobs, 1968, 1969).

In the case of a dust universe, the line element may be written in the form

$$ds^2 = dt^2 - (t^2-\mu^2a^{-2})^{2/3}\,[\xi^{c_1\mu^{-1}}dx^2+\xi^{c_2\mu^{-1}}dy^2+\xi^{c_3\mu^{-1}}dz^2]$$

$$\xi = (t-\mu a^{-1})(t+\mu a^{-1})^{-1};\ c_1+c_2+c_3 = 0$$

and the density and the shear are given by

$$6\pi\rho(t^2-\mu^2 a^{-2}) = 1$$

$$\sigma^2 = \frac{8}{3}\mu^2 a^{-2}(t^2-\mu^2 a^{-2})^{-2} .$$

Thus, near the singularity $t^2-\mu^2 a^{-2} = 0$, the shear dominates over the density so far as the dynamics of the model is concerned.

An interesting solution of this type which may be interpreted as due to matter with a directed flux of radiation is due to Raychaudhuri and Dutta (1974). The solution has been presented in the form

$$ds^2 = 2(\xi+\eta)^{2(\alpha+1)}d\xi d\eta - (\xi+\eta)^2(dx^2+dy^2)$$

with

$$T_{\mu\nu} = (p+\rho)v_\mu v_\nu - pg_{\mu\nu} + (T_{\mu\nu})_{\text{rad}}$$

$$8\pi p = 2(1+\alpha)(\xi+\eta)^{-2(2+\alpha)}$$

$$8\pi\rho = \frac{3+\alpha}{1+\alpha}\, 8\pi p .$$

$$v_1 = \left(\frac{\alpha+1}{\alpha+2}\right)^{\frac{1}{2}}(\xi+\eta)^{1+\alpha}$$

$$v_0 = \frac{1}{2}\left(\frac{\alpha+2}{\alpha+1}\right)^{\frac{1}{2}}(\xi+\eta)^{1+\alpha}$$

$$8\pi(T_{\mu\nu})_{\text{rad}} = \frac{\alpha(3\alpha+4)}{\alpha+1}(\xi+\eta)^{-2} \cdot \delta_{\mu 0}\delta_{\nu 0} .$$

α is an arbitrary constant which must be positive. However this makes $\rho/3 \leq p < \rho$.

That the above metric admits the Bianchi type I group of motions becomes evident if one makes the transformation

$$t = \xi+\eta, \quad r = \xi-\eta,$$

when the line element becomes

$$dS^2 = \tfrac{1}{2} t^{2(\alpha+1)}(dt^2 - dr^2) - t^2(dx^2+dy^2) \ .$$

The fluid velocity in this co-ordinate system has a non-vanishing space component, so that in the sense used by King and Ellis (1973) the metric represents a 'tilted universe'.

A solution admitting Bianchi type V was given by Schucking and Heckmann (1958) showing anisotropic expansion but no vorticity

$$dS^2 = dt^2 - R^2[dx^2 + S^2 e^{2x} dy^2 + S^{-2} e^x dz^2]$$

$$\dot{R}^2 = 1 + 2MR^{-1} + \tfrac{1}{3} a^2 R^{-4}$$

$$M = \tfrac{4}{3} \pi \rho R^3$$

$$S = \exp[a \int R^{-3} dt]$$

$$\sigma^2 = \tfrac{1}{3} a^2 R^{-6} \ .$$

More general models of the same group type have been considered by a number of Soviet investigators (Ruzmaikina and Ruzmaikin 1969; Grishchuk, Doroshkevich, and Novikov 1969). The line element is of the form

$$dS^2 = dt^2 - e^{2z}(a_1^2 dx^2 + a_2^2 dy^2) - a_3^2 dz^2 - 2a_{13} e^z dx dz$$

where the a's are functions of t alone. The vorticity ω vanishes if a_{13} vanishes. The approach to the singularity was specially studied. In all cases the density went to infinity and there were large density gradients in the three-spaces orthogonal to the matter velocity if $a_{13} \neq 0$.

A solution with expansion, shear, and vorticity has also been given by Batakis and Cohen (1975) and we reproduce below the solution due to Lukash (1974) which admits the Bianchi type VII group of motions. The metric is

$$ds^2 = dt^2 - g_{ik} dx^i dx^k \ .$$

$$g_{ik} = a^2 \begin{bmatrix} \cosh\mu + \sinh\mu . \cos 2kx^3 & \sinh\mu \sin 2kx^3 & -\nu e^{-\mu}\sin kx^3 \\ \sinh\mu \sin 2kx^3 & \cosh\mu - \sinh\mu \cos 2kx^3 & \nu e^{-\mu}\cos kx^3 \\ -\nu e^{-\mu}\sin kx^3 & \nu e^{-\mu}\cos kx^3 & C^2+\nu^2 e^{-\mu} \end{bmatrix}$$

where a, C, μ, and ν are undetermined functions of t. The solution is interpreted as representing a fluid with vorticity and associated with gravitational waves. The waves are circularly polarized and the wave vector k is along the x^3 axis and is a constant. The co-ordinate system is, in general, not co-moving and one has

$$T_i^0 = \frac{kR}{a^3 C}\left\{\cos kx^3 \, , \, \sin kx^3 , \, 0\right\}$$

where

$$R = -\frac{\nu}{2}\frac{a^3}{C} e^{-\mu} .$$

If $\nu = 0$, the motion is irrotational and if μ also vanishes the solution degenerates to a Bianchi type I universe without any gravitational wave. In particular if the universe is filled with a perfect fluid having $p = \alpha\rho$ where α is a constant, one has the integrals

$$kR = (p+\rho)a^4 C \, e^{\mu/2} \, v(1+v^2)^{\frac{1}{2}} = \text{const.}$$

$$\rho^{-\frac{1}{1+\alpha}} \, a^3 C(1+v^2)^{\frac{1}{2}} = \text{const.}$$

where $v = \beta(1-\beta^2)^{-\frac{1}{2}}$, β being the three-dimensional velocity.

The Bianchi type IX has the non-vanishing structure constants

$$C^1_{23} = C^2_{31} = C^3_{12} = 1.$$

The structure constants remain unchanged by a proper rotation of basis and thus, although the different directions in the three-space (i.e. the invariant varieties of the group) may not be equivalent metrically, they are equivalent topologically. Further the invariant varieties are all compact. (Indeed the isotropic universe with $k = +1$ is a special case admitting this type.) These, according to a fairly large number of cosmologists, are very desirable properties of the models that one attempts to build. Thus considerable attention has been given to a study of models admitting the type IX group and the nature of the singularity in these models has been extensively studied (Behr 1962, 1965; Shepley 1964; Misner and Taub 1969; Misner 1969b; Belinskii and Khalatnikov 1969; Belinskii, Khalatnikov, and Lifshitz 1970; Landau and Lifshitz 1971; Hawking 1969). We shall have occasion to refer in some detail to the interesting results regarding the nature of the singularity in a later chapter.

5.4. THE GÖDEL UNIVERSE

The Gödel universe (Gödel 1949, 1950) admits a group of motions of Bianchi type VIII having structure constants

$$C^1_{23} = -C^2_{31} = -C^3_{12} = 1 .$$

The invariant varieties for the group of motions are three-dimensional hypersurfaces containing null and time like lines (i.e. the orthogonal to these hypersurfaces are space like). However the Gödel universe admits a four-parameter Lie group of which the Bianchi type VIII is an invariant subgroup. The Gödel universe is thus both homogeneous and stationary. The stationary character as well as a breakdown of causality makes the model unsuitable for a representation of the observed universe.

Gödel gives the line element

$$\mathrm{d}s^2 = a^2[(\mathrm{d}x^0+\exp x^1\mathrm{d}x^2)^2 - \mathrm{d}x^{1^2} - \frac{\exp(2x^1)}{2}\mathrm{d}x^{2^2} - \mathrm{d}x^{3^2}] .$$

The signature condition is satisfied everywhere. The x^0 lines are the world lines of matter so that $v^\mu = (a^{-1},0,0,0)$ and the covariant components are $(a,0,a\exp x^1,0)$. Gödel showed that the Einstein field equations are satisfied if

$$8\pi\rho = a^{-2},\ \Lambda = -4\pi\rho,\ \omega^2 = \tfrac{1}{2}a^{-2} = 4\pi\rho = -\Lambda ,$$

where ω is the vorticity. It is noteworthy that while the Einstein static universe requires a positive cosmological constant to counter the gravitational attraction, Gödel has a negative Λ, as apparently the centrifugal repulsion due to vorticity dominates over the gravitational attraction. It is however permissible to consider the Gödel solution as one with $\Lambda = 0$ and a pressure equal to the energy density (the Zel'dovich limit).

The space time admits the following transformations:

(i) a translation along x^0 corresponding to the stationary property,

(ii) a translation along x^1,

(iii) a translation along x^3,

(iv) a translation along x^1 combined with a contraction along x^2: $x' = x^1 + b$, $x^{2\prime} = x^2 \exp(-b)$.

While the above establish the equivalence of all space-time points there is an additional rotational symmetry as may be exhibited by the transformation

$$\exp x^1 = \cosh 2r + \cos\phi \,\sinh 2r$$

$$x^2 \exp x^1 = \sqrt{2}\sin\phi\,\sinh 2r$$

$$\tan\left(\frac{\phi}{2} + \frac{x^0 - 2t}{2\sqrt{2}}\right) = \exp(-2r)\tan\frac{\phi}{2} \quad \text{where} \quad \left|\frac{x^0-2t}{2\sqrt{2}}\right| < \frac{\pi}{2}$$

$$x^3 = 2y .$$

The line element then assumes the form

$$ds^2 = 4a^2[dt^2 - dr^2 - dy^2 + (\sinh^4 r - \sinh^2 r)d\phi^2 + 2\sqrt{2}\sinh^2 r \, d\phi dt] \qquad (5.21)$$

which directly exhibits rotational symmetry about the y-axis.

ϕ is, however, a peculiar co-ordinate in the sense that it is space-like for small values of r, null for $r = \log(1+\surd 2)$, and time-like for greater values of r.

For $R > \log(1+\surd 2)$, the circle $r = R$, $t = y = 0$ (i.e. the ϕ line at $r = R$) is everywhere time-like. Such a time-like line indicates the possibility of returning back to a space-time point and thus one cannot assign any unique ordering in the time sequence of events. Hence we run into conflict with our idea of cause effect relationships.

Chandrasekhar and Wright (1961) have integrated the geodesic equations for the Gödel metric and have arrived at the result, amongst other things, that none of the closed time-like lines are geodetic. One may see that for the circular lines in a very simple manner. Using the line element (5.21) we see that for a ϕ line to be geodetic, we must have $g_{\phi\phi,1} = 0$ or either $r = 0$, or $\sinh r = \frac{1}{2}$. In either case the ϕ line is space like. Chandrasekhar and Wright therefore argued that a test particle cannot travel back into its past as envisaged by Gödel. However, the very existence of closed time-like lines indicate the possibility of pushing a particle back into the past by the application of suitable non-gravitational forces and thus involving a breakdown of causality.

The Gödel universe is not spatially closed and further has closed time-like lines; it has therefore been suggested that it should be 'excluded on physical grounds'. Also a non-coincidence of the rest frame of the cosmic matter with the inertial frame was apparently in conflict with the Mach principle. Gödel's solution thus gave an impetus for investigation of rotating universes. Ozsvath and Schucking (1969) exhibited a homogeneous rotating universe which has no closed time-like lines but the rotation is not rigid and the cosmological constant has to be retained. A non-vanishing cosmological term is present also in a stationary solution given by Ozsvath (1965). An inhomogeneous solution was given long ago by Van Stockum (1937) and rediscovered by Wright (1965). It did not require the cosmological term but the metric

$$ds^2 = dt^2 - \exp(-a^2r^2)(dr^2+dz^2) - (r^2-a^2r^4)d\phi^2 + 2ar^2 d\phi dt$$

contained closed time-like lines and had a singularity at a finite proper distance from the axis of cylindrical symmetry. More recently Maitra (1966) gave a non-homogeneous solution which was stationary, and with vanishing cosmological constant. The solution is regular everywhere, cylindrically symmetric, geodetically complete, and had no closed-time-like lines. Ozsvath (1966) has also given a stationary rotating solution with a distribution of dust and electromagnetic radiaton (see also Banerjee 1970). However all these solutions are unsuitable as models of the actual universe owing to their stationary character. They either do not give any spectral shift or a shift which is highly anisotropic with an 'average' value zero.

5.5. SPHERICALLY SYMMETRIC NON-HOMOGENEOUS SOLUTIONS

There have been a large number of investigations on spherically symmetric cosmological solutions which are not homogeneous; indeed homogeneity with spherical symmetry about any point would lead to the isotropic line element of Robertson and Walker. Again an inhomogeneous solution in the present discussion does not necessarily mean that the density distribution is inhomogeneous.

Bondi (1947), Landau and Lifshitz (1971), and Banerjee (1967), to name but a few, have given general discussions of spherically symmetric solutions, while such solutions with spatially constant density have been studied by Raychaudhuri (1955b), Thomson and Whitrow (1967, 1968), Bondi (1969), Banerjee (1970) etc. In view of the Misra-Srivastava theorem, discussed on p.83, these solutions must have isotropic expansion; however at the time of these investigations the theorem was not known and the isotropy of expansion was introduced as an additional simplifying assumption.

An interesting type of solution, which takes care of the discrete condensations that are found to occur in the actual universe, is due to Einstein and Strauss (1945) who considered an empty space region with a singularity at the centre and investigated whether the Schwarzschild field in this region could be fitted to an outside Robertson-Walker metric. It was found that the fit is possible if the material in the universe is pressure-free dust and the fit occurred at a constant co-ordinate radius of the cosmological field so that the boundary of the empty region is time dependent and the Schwarzschild mass of the condensation (i.e. the singularity) is given by

$$m = 4\pi\rho R^3 r_1^3 \left(1 + \frac{kr_1^2}{4}\right)^{-3}$$

where r_1 is the cosmological co-ordinate radius of the boundary of the two fields and is constant. ρ, R, and k have their usual significance for the outside universe.

The Einstein-Strauss fit depends critically on the vanishing of pressure in the universe and one may think that with a non-vanishing pressure the cosmic matter would be forced into the empty region introduced by Einstein and Strauss. In such cases an interesting solution was given by McVittie (1933). The condensation is again represented as a singularity and the line element is

$$ds^2 = e^\nu dt^2 - \frac{e^\lambda}{(1+kr^2/4)^2} (dr^2 + r^2 d\theta^2 + r^2 \sin^2\theta d\phi^2)$$

$$e^\nu = \left\{\frac{1-\mu/2r(1+kr^2/4)}{1+\mu/2r(1+kr^2/4)}\right\}^2$$

$$e^\lambda = R^2\left[1 + \frac{\mu}{2r}\left(1+\frac{kr^2}{4}\right)^{\frac{1}{2}}\right]^4$$

$$\mu = m/R$$

where m is a constant.

If one makes the substitution $r_1 = R.r$, the McVittie line element, at any particular instant, reduces to the Schwarzschild line element for small values of r, m being the mass. McVittie also showed that while in the cosmic co-ordinates the planetary orbits shrink and the 'mass of the condensation' μ decreases, in the Schwartzschild co-ordinates the orbits as well as the mass are independent of time.

The McVittie solution follows uniquely under the following conditions:

(i) The line element is spherically symmetric with a singularity at the centre.

(ii) The energy-stress tensor is that of a perfect fluid.

(iii) The fluid motion is shearfree.

(iv) The metric must asymptotically go over to the isotropic cosmological form.

Vaidya (1977) has recently given some metrics which apparently represent a rotating body in a static or expanding universe. However many features of the metrics seem so far obscure.

Amongst the host of other spherically symmetric solutions we may note particularly the investigation of Bonnor (1972) who has tried, with no apparent success, to construct a model to accommodate the variable cosmic density law $\rho \propto r^{-1\cdot 7}$ (given by de Vaucouleurs (1970, 1971) for our neighbourhood, (i.e. for r ranging from 10^7 to 10^{27} cm) and the observed red-shift luminosity relation. In a more recent paper Bonnor (1974) shows that the isotropic universe may evolve from a spherically symmetric universe in which the density on an initial space-like surface is an arbitrary function of the radial variable.

5.6. THE ROLE OF GRAVITATIONAL WAVES IN THE DYNAMICS OF ANISOTROPIC UNIVERSES

In our previous discussion on the 'missing mass' problem we referred to the possibility that part of this missing mass may be in the form of gravitational waves. However, as noted there gravitational waves, other than in the diffuse form, cannot be accommodated with the isotropic line element and, in any case, as gravitational waves do not contribute to the energy-momentum tensor $T^{\mu\nu}$, their role in the dynamics of cosmological models is somewhat obscure. Here we reproduce an argument of Hawking (1966) on the interaction of gravitational waves with matter. Hawking introduced a viscosity term $\lambda\sigma_{ab}$ in the expression for the energy-stress tensor where λ is the coefficient of viscosity and σ_{ab} is the shear. The divergence relation $T^{\mu\nu}{}_{;\nu}=0$ gives with $\mu = 0$

$$\frac{d\rho}{ds} + (p+\rho)\theta - 2\lambda\sigma^2 = 0$$

showing that the rest-mass energy of matter increases at a rate $2\lambda\sigma^2$ due to the absorption of the gravitational waves. (The term involving θ representing the usual expansion effect.)

The available energy in a gravitational wave decreases as R^{-4} in an expanding universe as for other zero rest-mass fields. In the early stages of the expanding universe, we may expect an equilibrium between electromagnetic and gravitational radiations with equal energy densities. Unless augmented by the conversion of matter, any primordial gravitational radiation would thus have a 'temperature' $\sim 2{\cdot}7$ K and would be too weak to be detected or have any appreciable effect on the dynamics of the universe. For incoherent gravitational waves, Hawking's perturbation calculation yields the result $\sigma^2 = 2\pi\rho_g$ where ρ_g is the 'energy density' of gravitational radiation and the Raychaudhuri equation (1955a) gives (with $\dot{v}_\mu = \omega_\mu = 0$)

$$\theta_{,\alpha}v^\alpha = -\tfrac{1}{3}\theta^2 - 4\pi\rho_g - 4\pi(\rho+3p)$$

showing that gravitational radiation has an attractive effect of the same magnitude as dust. The influence of gravitational radiation on the dynamics of the universe has been discussed in a different manner by Isaacson (1968) and Kafka (1970). (See also Isaacson and Winicour, 1973; Swinerd, 1977.)

5.7. THE LICHNEROWICZ UNIVERSES

In the so called Lichnerowicz universes, one considers an electrically charged fluid with infinite conductivity so that the electric field vanishes in the rest frame of the fluid but there is a non-vanishing magnetic field. The situation thus seems to be of only academic interest, yet as the solutions exhibit some novel features, it seems proper to make a mention of them. Ozsvath (1967) gave a solution in which the space time admitted a four-parameter simply transitive group of motions. Interesting stationary solutions in which there was a charged-dust distribution in rigid rotation were given by Som and Raychaudhuri (1968) and in particular a stationary equilibrium could be obtained with arbitrary values of the ratio of the charge and mass densities. Indeed the solutions were singularity free only if the charge density exceeded double the mass density. However there were closed time-like lines in all cases. De (1969) showed that in these 'universes' it is possible for test particles to describe closed time-like lines, unlike the case in the Gödel universe.

6
THE MICROWAVE RADIATION BACKGROUND

6.1. THERMAL NATURE OF THE RADIATION

A prediction about the existence of a universal radiation background is credited to Gamow and his school of workers on the so-called α-β-γ theory of synthesis of nuclear species. (For a review of this theory see Alpher and Herman (1950).) It was estimated that the temperature of this radiation, expected to be thermal, would be in the neighbourhood of 5 K. More than a decade after, Penzias and Wilson (1965) observed an excess temperature of 3·5 K for radiation of wavelength 7·4 cm. (Temperature of radiation of a particular wavelength signifies the temperature at which black-body radiation would have same intensity at that particular wavelength. Thus for a true thermal radiation the temperature would be the same at all wavelengths.) The excess temperature as observed by Penzias and Wilson was isotropic, unpolarized, and free from any significant fluctuation for the observed period which was about one year. Dicke, Peebles, Roll, and Wilkinson (1965) suggested an identification of the radiation with a remnant of the 'primordial fire ball' of big-bang cosmology. As one would then expect the radiation to be thermal, a very large number of subsequent observations were directed towards measuring the temperature of the radiation at different wavelengths. Table 6.1 shows some of these observational results. The table is representative rather than exhaustive.

The observations, by and large, support a black-body distribution at a temperature ~2·7 K in our locale. While there were some abnormally high temperatures observed by Shivanandan *et al.* (1968) and a line was apparently observed by Muehlner and Weiss (1970), later observations by these schools or by others have not confirmed the findings. However, the high flux, apparently isotropic, observed by Shivanandan *et al.* have not been clearly explained.

For 2·7 K, the crucial wavelengths for deciding whether the radiation has indeed the Planck distribution lie in the

TABLE 6.1

Observed temperature of the microwave background radiation at different wavelengths

Observer	Wavelength studied (cm)	Temperature (K)
Penzias and Wilson (1965)	7·4	3·5 ± 1·0
Roll and Wilkinson (1966)	3·2	3·0 ± 0·5
Howell and Shakeshaft (1966)	20·7	2·8 ± 0·6
Howell and Shakeshaft (1967)	73·5, 49·2	3·7 ± 1·2
Stokes, Partridge, and Wilkinson (1967)	3·2	2·69(-0·21 to +0·16)
	1·6	2·78(-0·17 to +0·12)
Boynton, Stokes and Wilkinson (1968)	0·86	2·56(-0·22 to +0·17)
	0·33	2·46(-0·44 to +0·40)
Welch, Keachie, Thornton, and Wrixon (1967)	1·50	2·0 ± 0·8
Ewing, Burke, and Staelin (1967)	0·92	3·16 ± 0·26
Puzanov, Salomonovich, and Starkovich (1967)	0·82	2·9 ± 0·7
Shivanandan, Houck, and Harwit (1968)	0·04-0·13	~8
Houck and Harwit (1969)	0·04-0·13	~6
Pipher, Houck, Jones, and Harwit (1970)	0·05-0·15	2·7 ± 0·1
Muchlner and Weiss (1970)	spectrum consistent with 2·7 K with a strong emission feature between 0·08 and 0·1 cm superposed.	

continued......

continued.......

Blair, Beery, Edeskutz, Hiebert, Shipley, and Williamson (1971)	0·008-0·6	3·1(-2·0 to +0·5)
Beery, Martiz, Nolt, and Wood (1971)	reports a line at 0·08 cm (atomspheric?)	
Mather, Werner, and Richards (1971)	no line at 0·08 cm	
Kislyakov, Chernyshev, Lebskii, Mal'tsev, and Serov (1971)	0·36	2·4 ± 0·7
Beckman, Ade, Huizinga, Robson, Vickers, and Harries (1972)	0·04-0·1	consistent with 3 K, no evidence of any line
Millia, McColl, Pederson, and Vernon (1971)	0·33	2·61 ± 0·25
Houck, Soifer, Harwit, and Pipher (1972)		2·7 ± 0·1
Muehlner and Weiss (1973a)	0·18-1·0	2·7(-0·2 to +0·4)
	0·13-1·0	2·8 ± 0·2
	0·09-1·0	≤ 2·7
	0·054-1·0	≤ 3·4
Williamson, Blair, Catlin, Hiebert, Loyd, and Romero (1973)	0·03-0·6	consistent with 2·7
Muehlner and Weiss (1973b)	0·086-1·0	2·55(-0·45 to +0·25)
	0·074-1·0	2·45(-1·05 to +0·45)
	0·054-1·0	2·75(-2·75 to +0·8)
Boynton and Stokes (1974)	0·3	2·48 ± 0·54
Woody, Mather, Nishioka, and Richards (1975)	0·06-0·25	$2{\cdot}99^{+0{\cdot}07}_{-0{\cdot}14}$
Grenier, Roucher, and Talureau (1976)	~ 0·02	< 3·4

submillimetre region. Unfortunately, in this region atmospheric emissions make the data highly suspect and that accounts for the very large uncertainties in this region (cf. Muehlner and Weiss 1973a,b) and paucity of data, especially below 0·08 cm.

6.2. THE UNIVERSALITY OF THE RADIATION

If this microwave background is indeed a relic of the primeval radiation, one would expect such a radiation in the interstellar space and also in other galaxies. The difficulties of obtaining evidence for the radiation, even if present in such places, are obvious, especially so far as other galaxies are concerned. There is some fairly convincing evidence of its existence in the interstellar region. In the spectra of some stars there are absorption lines attributable to cyanogen molecules (CN) in the interstellar space. One of these lines (λ = 3874·61 Å) is due to electronic excitation in non-rotating CN molecules while two other close lines (λ = 3874·00 and 3875·76 Å) are due to rotationally excited CN molecules. The energy of rotational excitation corresponds to a photon of wavelength 2·64 mm. Thus from a measurement of the relative intensities of these lines one can determine the 'rotational temperature' of the CN molecules and if the rotational excitation be due to radiation, this would give the temperature of 2·64 mm radiation. Some time before the discovery of the microwave background, McKellar (1941) from observations of the spectra of Zeta Ophiuchi, arrived at a temperature of ~2·3 K.

However there may be some doubt as to whether the excitation of CN molecules is due to a radiation field or to some other excitation mechanism such as collision with electrons or protons. In the latter case, the temperature deduced from CN excitation would give only an upper bound to the temperature of radiation at λ = 2·64 mm. To meet this criticism the CN temperature has been determined from spectra of stars in widely different regions so that the conditions along the line of sight may be expected to be significantly different. However the temperatures determined by taking different stars show remarkable agreement amongst themselves and also with the temperature of the radiation background determined from terrestrial observations. Table 6.2 shows the temperatures deduced from spectra of different stars (Bortolot, Clauser, and Thaddeus 1969; Sciama 1971; Hegye, Traub and Carleton 1974); Penzias, Jefferts, and Wilson 1972).

TABLE 6.2

Rotational temperature of CN molecules in interstellar space from intensity of absorption lines in stellar spectra

Star	Temperature (K)
Zeta Oph	$2{\cdot}74 \pm 0{\cdot}22$ $2{\cdot}9^{+0{\cdot}4}_{-0{\cdot}5}$ (for $\lambda = 1{\cdot}32$ mm)
Zeta Per	$2{\cdot}82 \pm 0{\cdot}30$
55 Cyg	$< 5{\cdot}5$
AE Aur	$3{\cdot}5 \pm 2{\cdot}3$
20 Aql	$2{\cdot}5 \pm 1{\cdot}8$
HD 12 953	$3{\cdot}7 \pm 0{\cdot}7$
13 Ceph	$2{\cdot}8 \pm 0{\cdot}4$
HD 26 571	~ 3
X Per	$2{\cdot}8 \pm 0{\cdot}8$
BD 66°1675	$2{\cdot}39 \pm 0{\cdot}4$
BD 66°1674	$2{\cdot}45 \pm 0{\cdot}6$

In a similar manner the absence of some absorption lines due to the second rotational state of CN (corresponding to excitation wavelength $\lambda = 1{\cdot}3$ mm) and the first rotational states of CH and CH^+ (excitation wavelengths 0·56 and 0·36 mm respectively) allow one to set upper bounds for the radiation temperatures at these wavelengths; these are found to be consistent with the 2·7 K background.

Is there any evidence for the existence of this radiation

outside our galaxy? In recent years several extragalactic X-ray sources have been discovered largely due to the satellite UHURU. Some of these are diffuse extended sources with sizes ~10 pc and are associated with different clusters. These X-rays may be generated either by thermal bremsstrahlung or by inverse Compton process (Brecher and Burbidge 1972; Miley, Perola, Van der Kruit and Van der Laar 1972; Solinger and Tucker 1972; Gursky, Solinger, Kellog, Murray, Tananbaum, Giacconi, and Cavaliere 1972; Bowles, Patrick, Sheather, and Eiband 1974). The data from Copernicus seem to indicate that the thermal-bremsstrahlung origin is more likely (Griffiths and Peacock 1974). The data from the Coma cluster fits in with the thermal-bremsstrahlung theory if one takes the temperature of the plasma as kT ~8 keV (Gorenstein, Bjorkholm, Harris, and Harnden 1973). Such a plasma, interacting with the microwave photons via the Compton scattering, would cause a decrease of the temperature of the radiation (Zel'dovich and Sunyaev 1970). Such a decrease has apparently been observed in the direction of the Coma cluster of magnitude $\sim 10^{-3} - 10^{-4}$ K (Parijskij 1972). See also Gull and Northover (1976); Lake and Partridge (1977) however report an increase of temperature in the direction of the Coma cluster.

6.3. THE ISOTROPY OF THE RADIATION

The isotropy of the background radiation is an important subject of investigation for a variety of reasons and a fairly large number of observations have been made both on small- and large-scale anisotropies. It is unnecessary for us to go into the details of the technical arrangements and the results that have been presented. Let us merely emphasize that no significant anisotropy has been observed either on a small or large scale and the limit to the anisotropy may be expressed in the form $\Delta T/T \leq 0{\cdot}001$ or $< 0{\cdot}1$ per cent. (Conklin and Bracewell 1967a,b; Partridge and Wilkinson 1967; Conklin 1969; Penzias, Schraml, and Wilson 1969; Boynton and Partridge 1973; Parijskij 1973a; Caderni, de Cosmo, Fabbri, Melchiorri, B.F. and Natale 1977.)

The isotropy of the microwave background raises an interesting possibility. It allows us to define a rest frame and consequently a velocity in an 'absolute' sense. This is because a radiation field which appears isotropic to an observer would not appear so to a relatively moving observer - to the latter the radiation is still black but its temperature is different in different directions being given by

$$T'_\theta = T(1+z_\theta)^{-1} , \tag{6.1}$$

where θ is the angle between the direction of observation and the direction of the velocity $\underline{v}$ and the relativistic Doppler-shift is given by

$$\frac{\nu'_\theta}{\nu} \equiv (1+z_\theta)^{-1} = \frac{1-v\ \cos\theta}{(1-v^2)^{\frac{1}{2}}} \quad . \tag{6.2}$$

The formula (6.2) can be obtained directly from the Lorentz transformation (Weinberg 1972b) but may be shown to hold good for more general cases where the frequency change may be for reasons other than Doppler effect (Ellis 1971).

It is true that this possibility of defining an absolute frame of rest is implicit in the assumption of homogeneity of the universe in cases of non-static models because relatively moving observers make different space sections from the four-dimensional manifold. However, before the discovery of the microwave background, the homogeneity postulate appeared merely as a useful working approximation and operationally it was hardly possible to define the frame of homogeneity due to the quite apparent irregularities both in the galactic distributions and in the observed red shifts. With the high degree of isotropy of the microwave background one can define the absolute velocity with a fair accuracy. Of course the velocity to be measured must be sufficient to produce a measurable change in the temperature of the background radiation. However the introduction of an absolute velocity and rest frame must not contradict the special theory of relativity; hence the rest frame is in no way a preferred frame so far as the basic laws of nature are concerned (cf. Bondi 1962b).

An attempt to determine the velocity of the earth relative to this absolute frame has been made from an analysis of the observed anisotropy of the microwave background. Thus Henry (1971) carried out balloon observations at an altitude of 24 km and found a velocity of 320 ± 80 km s^{-1} in a direction $10\frac{1}{2}$ ± 4 hr right ascension, -30 ± 25° declination. (The temperature anisotropy observed had an amplitude 3·2 ± 0·8 mK.) The effect of galactic radiation was estimated and corrected for by assuming a spectral index of -2·80 and scaling a 404 MHz map of the galaxy (Pauliny-Toth and Shakeshaft 1962) to 10·15 GHz (Henry's observation band was 9·9 to 10·4 GHz). Astronomical data lead to a velocity of the earth relative to the centre of the supercluster, of which our galaxy is a member, of about 400 km s^{-1} in the direction 14 hr,-20°. Henry considers that a reasonable estimate of the uncertainty in the astronomical value is ± 200 km s^{-1}, ± 2hr, ± 20°. Thus, within the range of uncertainty, which is unfortunately rather large, there is no relative motion between the local supercluster and the frame of isotropy of the background radiation. More recently Corey and Wilkinson (1976) found the velocity of the earth to be 270 ± 70 km s^{-1} in a direction 13 h ± 2h, -25° ± 20° from an analysis of balloon observations at 19 GHz.

Recently Rubin *et al.* (1976) have estimated from the observed anisotropy of red shifts the velocity of the galaxy. The magnitude is about 500 km s^{-1} and the direction also disagrees with that arrived at from the microwave radiation anisotropy observations.

Hawking (1969) on the other hand, has tried to set an upper bound to the vorticity of the homogeneous cosmological models by considering the small anisotropy of the background radiation to be due to vorticity and shear of the cosmic velocity field rather than to any peculiar velocity that the earth may have. We now have (cf. equation 6.1)

$$T_R = T_E(1+z)^{-1} \tag{6.3}$$

where T_R is the temperature of the radiation as observed by the receiver and T_E is the temperature of the last scattering surface. The red-shift z depends on expansion only in the isotropic universe but would depend, in general, also on the velocities arising from shear and vorticity. z would then be angle dependent, and a general formula is

$$z = (k^{\mu}v_{\mu})_{E}(k^{\mu}v_{\mu})_{R}^{-1} - 1 \qquad (6.4)$$

where v^{μ} is the velocity vector of the cosmic matter and k^{μ} the null vector for the light ray. The subscripts E and R indicate the values for the emitter and receiver respectively. Thus the observed anisotropy in the temperature of the radiation is linked with v^{μ} and hence may be used to set limits on the values of shear and vorticity. Considering homogeneous but anisotropic universes Hawking derives a formula for $(1+z)$ correct to the first order; the formula is then applied to Bianchi types I, V, VII_0, VII_h and IX by Collins and Hawking (1973b). The particular Bianchi types were selected as they admit the Robertson-Walker metric. (For a discussion of the Bianchi types, see Ellis and MacCallum (1969).) For the closed universes (i.e. Bianchi type IX) it appears that ω/H (if the missing matter be in the form of ionized hydrogen) is less than $1{\cdot}5 \times 10^{-8}$ and is less than 10^{-11} if there has been no scattering since the decoupling of matter and radiation at $z \sim 1000$.

For the open models vorticity is *a priori* excluded, for the type I and for the other types a limit on ω/H of about 10^{-4} may be set. These very low values of the upper limit to vorticity give support to the Mach principle.

Two remarks regarding the Collins-Hawking calculation seem pertinent. It is not clear how the limits would be changed if one admits some departure from homogeneity. Again, for the anisotropy of the microwave background of large angular scale they take $(\Delta T/T)_{max} \le 10^{-3}$, a value given by Partridge and Wilkinson (1967). However their observations extended only over certain areas of the sky and may not hold for the entire celestial sphere.

From an observation covering approximately half of the celestial sphere, the soft X-ray background at ~10 keV is found to be remarkably isotropic - the upper limit for any 12 or 24-hour variation of intensity being 1 per cent (Schwartz 1970). Assuming the radiation to be cosmological, it is possible to set limits on present-day shear and vorticity. This, in principle, is superior to that based on the microwave background, for while the data for the latter covers only two circles of the celestial sphere, the X-ray studies are for approximately half of the sky (Wolfe 1970).

Limits to the shear have also been studied on the basis of the limit to departure from the Planck distribution by taking the Bianchi type I model and including the effects of neutrinos on the dynamics of the universe (Rasband 1971). However at the present level of accuracy of observations, the limit to the value of the shear set by the isotropy of the background cannot be narrowed down by this consideration.

For the open models Collins and Hawking obtained the interesting result that the radiation temperature should have extremely high values (indeed, in their calculation, infinite) in a particular direction. While such a 'hot spot' would give observational evidence in favour of an open model, as yet the inadequate observational data cannot be held as conclusive evidence against the existence of such a hot spot.

Hawking and Ellis (1968) have pointed out that if we assume that the isotropy of the microwave background at our locale is exact and believe also in the Copernican principle that ours is no exceptional position so that the isotropy of the microwave background obtains everywhere, then the universe must have the Robertson-Walker metric. The isotropy of the radiation field defines a unique velocity field (representing say the world line of the galaxies) which may be assumed to be geodetic. It now follows from the relativistic Liouville equation applied to collisionless photons that the velocity congruence is shear free and normal (Tauber and Weinberg 1961). The Einstein equations of general relativity now lead to the conclusion that the universe is isotropic and homogeneous (Raychaudhuri 1955a). The same conclusion has been arrived at in a somewhat different manner by Ehlers, Geren, and Sachs (1968).

The isotropy on a small angular scale -$\Delta T \leq 0{\cdot}0043$ at $\lambda = 0{\cdot}35$ cm (Boynton and Partridge 1973) and $\Delta T \leq 0{\cdot}0008$ at $\lambda = 2{\cdot}8$ cm (Parijskij 1973b) allows one to set limits on the irregularities in the distribution of matter which the radiation has traversed in reaching us. This problem has been extensively studied (Sachs and Wolfe 1967; Rees and Sciama 1968; Wolfe 1969; Dautcourt 1969; Longair and Sunyaev 1969; Chibisov and Ozernoy 1969). Sachs and Wolfe explicitly integrated the equations of linearized perturbations in the case where the background is the Einstin-de Sitter universe (i.e. flat-space Robertson-Walker metric) and $p = 0$ or $\rho/3$. In the perturbed universe, the equations of the null geodesics were integrated. The anisotropy of z and consequently of the temperature of the microwave background is then estimated assuming the radiation to be cosmological. Sachs and Wolfe found a density fluctuation now of order 10 per cent with characteristic length of order 1000 Mpc would cause anisotropies of order 1 per cent in the temperature. (The Sachs-Wolfe value would be reduced by a factor of 2 if, instead of their H_0, we use $H_0 = 50$ km s^{-1} Mpc^{-1}.) Rees and Sciama find that the effect on the temperature due to an irregularity in the form of a condensation in a homogeneous background (cf. the Einstein-Strauss solution discussed in Chapter 5) is quite complicated and may be of either sign.

6.4. ALTERNATIVE IDEAS ABOUT THE ORIGIN OF THE BACKGROUND RADIATION

The existence of a hot, dense plasma interacting strongly with radiation accounts quite naturally for the Planck distribution and one can look upon the radiation as a remnant of the primeval radiation at the big bang. In the alternative steady-state theories, one must not only find out some source of the radiation, but must also explain the mechanism of thermalization and isotropization. As so far it has not been possible to put forward very credible theories on these lines, most cosmologists take this radiation as a strong evidence in favour of big-bang models. Nevertheless we shall give a short account of the alternative ideas that have been occasionally advanced. Thus Hoyle and Wickramsinghe (1967) and Narlikar and Wickramsinghe (1968) have suggested that the radiation emitted in the optical region by stars may be absorbed by interstellar grains which, in their turn, re-emit the energy in the microwave region. Again it has been supposed that a very large number of discrete sources may co-operate to build up a nearly isotropic background (Gold and Pacini 1968; Wolfe and Burbidge 1969; Wagoner 1969). However it is difficult to see how the spectral distribution would be Planckian in these cases and for discrete sources, their number N must be almost as great as the number of galaxies in order that the expected fluctuation $\sqrt{N}/N$ may not contradict with the very low limits set by observations. (For a recent theory of discrete source origin of the microwave background, see Rowan-Robinson 1974.)

Nevertheless, besides these protagonists of the steady-state theories, there are others who consider that a 'cold' big-bang offers some advantages over a hot big-bang so far as the formation of condensations like the observed galaxies are concerned. Thus Zel'dovich (1963) conjectured that a uniform distribution of hydrogen at ~ 0 K undergoes a phase transition to molecular solid-state when the density drops to 1 g cm^{-3} and would subsequently break up into fragments of planetary mass. Layzer and Hively (1973) suggested that the medium may initially solidify in the metallic state and shatter before undergoing a further phase transition to the molecular solid-state. In this way an explanation of galaxy formation was imagined to be possible. However with a cold big bang, not only the thermalized radiation background but also the observed abundance of deuteron and helium would be difficult to understand.

Layzer and Hively (1973) put forward the hypothesis that the microwave background is due to discrete sources and the radiation has been thermalized by interaction with dust grains. The observed characteristics (i.e. the temperature of 2·7 K, the black-body distribution, and isotropy) of the radiation can be attained if most of the matter in the early universe condensed into stars in the mass range 5-10 M. These stars, it was further conjectured, have released most of their energy at red shifts $25 \leq z \leq 50$ and have also ejected the heavy elements which might have gone to form the grains which are imagined to be responsible for the thermalization of the radiation.

The grains postulated would affect the magnitudes and

colours of distant objects and it was suggested that the then failure to observe quasars with $z \gtrsim 3$ was due to extinction by these grains. However, shortly after Layzer and Hively's work, quasars with z as high as 3·5 were observed (Carswell and Strittmatter 1973; Wampler, Robinson, Baldwin, and Burbidge, 1973). A modified version of the Layzer-Hively theory has since then been given by Wickramsinghe, Edmunds, Chitre, Narlikar, and Ramadurai (1975) where the thermalization is due to graphite whiskers. Because of their peculiar shape, the absorption and consequent extinction is drastically reduced.

Carr (1975) has investigated the possibility that the microwave background may be due to photons emitted from the primordial black holes by the Hawking process in an initially cold universe.

While the common idea is that the background radiation is a relic of the radiation that existed right from the beginning (i.e. the big-bang singularity) Rees (1971, 1972) points out that for thermalization it is only necessary that the radiation has existed in the highly dense plasma regime corresponding to $z \gtrsim 1500$. However, if the radiation originated at a red shift z_c when the temperature was $2{\cdot}7\,(1+z_c)$, the plasma at that temperature should be able, within the time available, to generate enough photons to produce a radiation field with full black-body intensity. This leads to a value for the thermalization red shift as

$$(1+z_m) \approx 4 \times 10^5\, \sigma^{-3/2}\, f^{-1} ,$$

where σ is the density parameter and f is a measure of the 'clumpiness':

$$f = \langle \rho_m^2 \rangle | \langle \rho_m \rangle^2 \geq 1,$$

ρ_m being the density of matter. It was thus concluded that if the early universe possessed the maximum degree of irregularity compatible with the present large scale homogeneity and isotropy, dissipation of energy at $z \gtrsim 10^4$ could have generated a thermal radiation-background with present temperature ~3 K.

Rees (1968) has also pointed out that while the observed isotropy of the microwave background indicates an isotropy up to the last scattering surface, anisotropic expansion prior to that epoch would show up as a linear polarization of the radiation and a distortion of the spectral distribution from the Planck form. Rees finds that in case of models with axial symmetry, the polarization ε would be given by

$$\varepsilon \sim (0{\cdot}1 \text{ to } 3) \times \text{anisotropy}.$$

No polarization (up to a few parts in thousand) has been observed (Nanos 1973). Brans (1975) has considered the effect of the metric on the propagation of polarized radiation and finds that a general anisotropic metric rotates the plane of polarization - the amount of rotation depending on the length of the path. Thus if the radiation is coming from sources at different distances, the absence of polarization would not be a convincing argument in favour of early isotropy.

The problem of the origin of the microwave background may be looked upon as a problem of entropy production. The entropy S of the radiation is conserved in purely adiabatic expansion and thus the non-dimensional number S/kn where k is the Boltzmann constant and n the baryon number is a conserved quantity, except for irreversible processes. With $T \sim 2{\cdot}7$ K and $\rho_m \sim 10^{-30}$ g cm^{-3}, this number has a value $\sim 10^9$. One may assume that this is a number fixed right from the beginning of the universe and indeed this is usually done. However an alternative approach is to start with a vanishingly small value for this number and try to explain it in terms of dissipative processes. Local irreversible changes, such as those taking place in stars, cannot explain this high entropy - if one takes the stellar temperatures as $\sim 10^7$ K the entropy per baryon is smaller by several orders of magnitude. In the cosmological scale, if one takes the Robertson-Walker line element, then the only dissipative term possible in the energy-stress tensor is that due to bulk viscosity (see Section 3.3). However Weinberg (1971) has shown that this cannot account for the present high value of the entropy per baryon. One is thus led to the conclusion that if there has been the requisite entropy generation, the early universe must have departed from isotropy and homogeneity.

A novel approach in this problem is due to Zel'dovich (1972a, 1973). He assumes that the universe near the singularity was filled with cold baryons and that over an average isotropic and homogeneous metric there was superposed metric fluctuations of amplitude $\sim 10^{-4}$. Then one could have the observed ratio of photons and baryons as also an understanding of the mass and density of galactic clusters. However the equation of state $p = \rho$ for baryons was assumed and the magnitude of the perturbation was introduced *ad hoc*. In the Zel'dovich model the transition to the 'hot' universe occurred at $t \ll \hbar/mc$ so that the nucleogenesis calculations are not affected.

6.5. INTERACTION OF HIGH ENERGY PARTICLES WITH THE MICROWAVE PHOTONS

6.5.1. *Elastic scattering* $e+\gamma \rightarrow e+\gamma$

The elastic scattering of photons by free electrons - the Compton scattering - results in a degradation of the energy of the photon which is gained by the electron. However, here we are concerned with the case where in the observers' system the electron energy is high and the photon energy is low. In this case the electron energy is quickly lost while the increase in photon energy may lead to the generation of X-rays and γ-rays from the original microwave radiation. This process, called the inverse Compton effect, has already been referred to as a possible explanation of the X-ray background. For ultrarelativistic electrons having energy $E = \gamma mc^2$ with $\gamma \gg 1$, the photons, after scattering through an angle θ (in the electron frame of reference), have an energy in the laboratory frame (Greisen 1971)

$$W \approx \frac{4}{3}\gamma^2 W_0(1-\cos\theta)\left[1 + \frac{4}{3}\frac{\gamma W_0}{mc^2}(1-\cos\theta)\right]^{-1}$$

where W_0 is the average energy of an initially isotropic flux of photons and is equal to $2{\cdot}7\ kT$ for black-body radiation at temperature T. With $T = 2{\cdot}7$ K and $\gamma < 6{\times}10^8$, the second term within the bracket may be neglected. Under this condition the distribution is symmetric around $\theta = 90°$ and $\langle W \rangle \approx 4\gamma^2 W_0/3 \approx 3{\cdot}6\gamma^2 kT$. Thus electrons of 1 Gev would scatter the photons of the microwave background as X-rays of about 3 keV energy while electrons of energy 10^{12} eV, as may be present in the Crab, would generate γ-ray photons of energy 3 GeV. How far this inverse Compton process contributes to the diffuse soft X-ray background is still a debated question.

The generation of energetic photons by the inverse Compton process is associated with an energy loss of the electrons. Using the expression $\sigma = 8\pi/3(e^2/mc^2)^2$ for the total scattering cross-section for this process, the mean free path of an electron is $(n\sigma)^{-1} = (u\sigma W_0^{-1})^{-1}$ where n is the number density of photons and u the energy density of the radiation field. As the average energy loss per collision is $4\gamma^2 W_0/3$, we get for the rate of loss of energy of the electrons

$$\frac{dE}{dt} = -\frac{32}{9}\pi\left(\frac{e^2}{mc^2}\right)^2 \gamma^2 c.u.$$

For the 2·7 K radiation field, $u = 0{\cdot}25$ eV cm^{-3} and if E is expressed in GeV, the above expression gives for the energy loss per year

$$\frac{dE}{dt} \approx -8 \times 10^{-10}\ E^2 .$$

On integration, we find that the time for the energy of the electron to be reduced to half its initial value (conveniently called the life time of an electron) is $\sim 2{\cdot}5 \times 10^6$ years for $E = 500$ GeV. Thus if the high-energy electrons are coming from outside our galaxy, there should be a decrease in the flux of electrons at energies above 100 GeV. Indeed some observers have claimed to observe a 'break' in the electron spectrum. Thus Matsuo, Mikuno, Nishimura, Niu, and Taira (1971) reported a steepening of the spectrum at 50 GeV while Anand, Daniel, and Stephens

(1971) found a similar effect at 150 GeV. However many others (Rubtsov and Zatsepin 1968; Bleeker, Burger, Debrenberg, Scheepmaker, Swanenberg, and Tanaka 1968; Danjo, Hayakawa, Makino and Tanaka, 1968; Anand, Daniel, and Stephens 1968; Rockstroh and Weber 1969) have found that in the region from 3 to at least 300 Gev, the electron flux closely follows a simple power-law with the spectral index -2·6. Very recently Muller and Meyer studied the spectrum in the region between 10 and 900 GeV and up to around 250 GeV, their data were well represented by a spectral index 2·66 ± 0·1. Even up to 900 GeV no change in the spectral index could be detected with certainty (Muller and Meyer 1973).

6.5.2. Interaction with protons

The microwave background contains a copious supply of photons ($\sim 400\ cm^{-3}$) of low energy and the threshold energy of protons for production of pions by interaction with these photons is $\sim 10^{20}$ eV. The effect has been investigated by a number of workers and cross sections have been calculated for different proton energies (Greisen 1966; Zatsepin and Kuzmin 1966; Stecker 1969a,b). The pions that are produced give rise to neutrinos, electrons, and photons. Again each time a pion is produced, the energy of the proton would be reduced by about 1/5th of its value. The absorption time for a proton comes out as $\sim 10^{16}$s for energy 10^{20} eV and has a minimum of 10^{15}s for proton energy of 10^{21}eV. In the earlier epochs, when the background radiation temperature was higher, the threshold energy and the absorption times were significantly lower. Thus one expects a sharp cut-off for protons of energies above 10^{21} eV unless the protons are coming from close neighbouring galaxies. (Protons of energies above 10^{18} eV are unlikely to be galactic as they cannot be trapped within the galaxy by the available magnetic fields $\sim 10^{-6}$ G.)

Lower energy protons (energy $\gtrsim 10^{18}$ eV) may produce electron-positron pairs by interaction with the background radiation and the energy of these pairs would be $\sim 10^{15}$ eV. This energy loss is small for protons but the pairs produced may give rise to high-energy photons by the inverse Compton process with the microwave background radiation. These photons again can interact with the background photons to produce new energetic electron-pairs and thus generate cascade showers. (The process $\gamma+\gamma \rightarrow e^{+}+e^{-}$ can occur if the product of the photon energies exceeds m^2c^4). We can depict the process as below:

Energetic protons (> 10^{18} eV)	⟶ interaction with background photons	Electrons (+ and -)
Electron pairs (~10^{15} eV)	⟶ inverse Compton effect	Photons of energy 10^{14} eV
Photons (high energy)	⟶ interaction with background photons	Electrons of high energy

6.5.3. The opacity of the universe to electromagnetic radiation

The threshold energy for an incident photon to produce an electron pair by interaction with the photons of the background radiation is ~$2{\cdot}5 \times 10^{14}$ eV and the cross section for the process has a maximum at about ~10^{15} eV. At this energy the mean free path of the photon is about 2×10^{22} cm which is smaller than the galactic dimensions. However the background radiation spectrum has a finite width and, as the energy of the incident photons increase, the decrease in the cross section is compensated by the increase in the number of photons (of lower energy) in the background which may now be effective in the pair production process. In this way the universe maintains a high opacity for all photons of energy above 10^{14} eV. Thus we may expect a cut-off of photons above this energy. (Jelley 1966; Gould and Schreder 1966, 1967; Sciama 1971).

6.6 THE ISOTROPY OF THE BACKGROUND RADIATION, PARTICLE HORIZONS, AND CAUSALITY

If in the universe with H_0 = 50 km s^{-1} Mpc^{-1} and q_0 = +1, the matter present is primarily in the form of ionized hydrogen (possibilities previously held to be likely but now almost excluded by observations) then the last scattering surface for the microwave background is at $z \approx 8{\cdot}6$ (i.e. the optical depth is unity at this z (cf. equation (4.34) p.76). If, however, the universe is of low density with $q_0 \approx 0$, (as is being favoured at present) the last scattering surface would be as early as the epoch of recombination of protons and electrons to form hydrogen atoms when the matter and radiation became decoupled ($z \sim 1000$). If the observations of isotropy of the background radiation extended over the entire sky, then we could have concluded that there was an uniformity extending over the entire last scattering surface. However, as the observations stand, we can legitimately say that the uniformity extended over a fairly large angular dimension of the last scattering surface. It would be our aim in this section to investigate whether these regions of uniformity were within their particle horizons at the epoch of scattering - if not, it would raise the question as to how there could be similar conditions at different regions which had no possibility of intercommunication?

The discussion would necessitate a calculation of the 'distance' between two points having the same value of r, the radial co-ordinate, but different values of θ. We perform the calculation following a method given by Tolman (1934G). Tolman argues that as all positions in the universe are equivalent, it must be possible to transfer the origin to any desired point, still retaining the form of the Robertson-Walker line element. This he does by considering the embedding of the four-dimensional space time in a five-dimensional manifold. Thus with the transformation

$$\bar{r} = r(1+kr^2/4)^{-1} \tag{6.5}$$

the Robertson-Walker line element assumes the form

$$ds^2 = -R^2\left(\frac{d\bar{r}^2}{1-k\bar{r}^2} + \bar{r}^2 d\theta^2 + \bar{r}^2 \sin^2\theta d\phi^2\right) + dt^2 \quad . \tag{6.6}$$

Now, introducing four variables (not independent) in place of $\bar{r}, \theta, \phi$ by the relations

$$z_1 = (1-k\bar{r}^2)^{\frac{1}{2}}, \; z_2 = \bar{r}\sin\theta.\cos\phi, \; z_3 = \bar{r}\sin\theta\sin\phi$$

$$z_4 = \bar{r}\cos\theta, \quad \text{i.e. } \bar{r}^2 = z_2^2 + z_3^2 + z_4^2 \tag{6.7}$$

so that $z_1^2+k(z_2^2+z_3^2+z_4^2) = 1$, the line element becomes

$$ds^2 = -R^2(dz_1^2+dz_2^2+dz_3^2+dz_4^2)+dt^2 . \tag{6.8}$$

For the two points $(\bar{r}_1,0,0)$ and $(\bar{r}_2,\theta,0)$, the z values are respectively

$$\begin{aligned}(\bar{r}_1,0,0) &\to (1-k\bar{r}_1^2)^{\frac{1}{2}},\ 0,0,\bar{r}_1\\ (\underline{r}_2,\theta,0) &\to (1-k\bar{r}_2^2)^{\frac{1}{2}},\ \ \bar{r}_2\sin\theta,\ 0,\ \bar{r}_2\cos\theta .\end{aligned} \tag{6.9}$$

By a transformation in the $z_1\ z_4$ plane of the form

$$\left.\begin{aligned} z_1' &= z_1\cos\alpha + z_4\sin\alpha\\ z_4' &= -z_1\sin\alpha + z_4\cos\alpha\end{aligned}\right\}\ \sin\alpha = \bar{r}_1,\quad k = +1 \tag{6.10a}$$

$$\left.\begin{aligned} z_1' &= z_1\cosh\alpha + z_4\sinh\alpha\\ z_4' &= z_1\sinh\alpha + z_4\cosh\alpha\end{aligned}\right\}\sinh\alpha = +\bar{r}_1,\quad k = -1 \tag{6.10b}$$

we get for all k

$$(\bar{r}_1,0,0) \to 1,\ 0,\ 0,\ 0$$

$$(\bar{r}_2,\theta,0) \to (1-k\bar{r}_{12}^2)^{\frac{1}{2}},\ \bar{r}_2\sin\theta,0,\ (\bar{r}_{12}^2-\bar{r}_2^2\sin^2\theta)^{\frac{1}{2}}$$

where

$$\bar{r}_{12}^2 = \bar{r}_1^2+\bar{r}_2^2-\bar{r}_2^2\ \bar{r}_1^2\ (1+\cos^2\theta)\ -2(1-\bar{r}_2^2)^{\frac{1}{2}}(1-\bar{r}_1^2)^{\frac{1}{2}}\bar{r}_1\bar{r}_2\cos\theta$$

(for $k = +1$)

$$= \bar{r}_1^2+\bar{r}_2^2+\bar{r}_2^2\ \bar{r}_1^2\ (1+\cos^2\theta)\ -2(1+\bar{r}_2^2)^{\frac{1}{2}}(1+\bar{r}_1^2)^{\frac{1}{2}}\ \bar{r}_1\ \bar{r}_2\cos\theta$$

(for $k = -1$) .

Switching back to a new (r',θ',ϕ') co-ordinates defined by

$$z_1' = (1-kr'^2)^{\frac{1}{2}},\quad z_2 = r'\sin\theta'\cos\phi',\quad z_3 = r'\sin\theta'\sin\phi',$$
$$z_4' = r'\cos\theta'$$

we get back the line element (6.6) in these new co-ordinates and the co-ordinates of the points become

$$(\bar{r}_1,0,0) \to (0,\ldots,\ldots)$$

$$(\bar{r}_2,\theta,0) \to (\bar{r}_{12},\sin^{-1}[\bar{r}_2\,\sin\theta/\bar{r}_{12}],0)\ .$$

Thus the origin has shifted to the first point, and the radial co-ordinate of the second point is given by $\bar{r}_{12}$. If the two points are on the last scattering surface $\bar{r}_1 = \bar{r}_2$ and then

$$\bar{r}_{12} = 2\bar{r}_1\sin(\theta/2)[1-k\bar{r}_1^2\,\sin^2\theta/2]^{\frac{1}{2}} \quad \text{for all } k. \tag{6.11}$$

The points will lie within or beyond their particle horizons according as

$$\chi_{12} \equiv \int_0^{\bar{r}_{12}} \frac{dr}{(1-kr^2)^{\frac{1}{2}}} \tag{6.12}$$

is less or greater than $\int_0^t R^{-1}dt$ where t is the epoch of emission of the light from the last scattering surface that we are receiving now. $t=0$ corresponds as before to the epoch $R=0$ $(z\to\infty)$. Now

$$\chi_{12} = \left.\begin{array}{lll} \sin^{-1}\bar{r}_{12} & 2\sin^{-1}(\bar{r}_1\,\sin\theta/2) & k=+1 \\ \bar{r}_{12} \quad = & 2\,\bar{r}_1\,\sin\,\theta/2 & k=0 \\ \sinh^{-1}\bar{r}_{12} & 2\sinh^{-1}(\bar{r}_1\sin\theta/2) & k=-1 \end{array}\right\} \tag{6.13}$$

Using equations (6.5), (4.2) (p.49), and (4.29) (p. 61) we get:

$$\bar{r}_1 = R_0^{-1}H_0^{-1}q_0^{-2}(1+z)^{-1}\left\{q_0 z+(q_0-1)\left[(1+2q_0 z)^{\frac{1}{2}}-1\right]\right\} . \qquad (6.14)$$

Also from equations (4.14), (4.25), (4.26), and (4.27) we have, remembering that $z\to\infty$ as $t\to 0$:

$$\left.\begin{aligned}\int_0^t R^{-1}\mathrm{d}t &= \cos^{-1}\left[1-\frac{(2q_0-1)}{q_0(1+z)}\right] && k=1\\ &= 2H_0^{-1}R_0^{-1}\,(1+z)^{-\frac{1}{2}} && k=0\\ &= \cosh^{-1}\left[\frac{1-2q_0}{q_0(1+z)}+1\right] && k=-1\ .\end{aligned}\right\} \qquad (6.15)$$

Using z for the last scattering surface as 8·6 (k = +1, q_0 = +1) one gets from equations (6.13), (6.14), and (6.15) that points separated by more than about 30° would be beyond their horizons. However observations do show an uniformity extending over larger angular separations. (In case z is very large ~1000, almost any finite angular separation would make the points beyond their horizons.)

Misner (1969b) sought to remove this difficulty with the supposition that the universe was highly anisotropic in its early stages. Misner noted that for the Kasner empty universe with the line element

$$\mathrm{d}s^2 = \mathrm{d}t^2 - t^2\mathrm{d}x^2 - \mathrm{d}y^2 - \mathrm{d}z^2$$

there exists no horizon for propagation of light in the x-direction. For the anisotropic universe, admitting the Bianchi type IX group of motions it was found that near the singular state, the universe approaches closely the Kasner model, abolishing horizons in a particular direction. However this direction does not remain fixed but makes a type of oscillatory change so that there is effectively an absence of horizon in any direction whatsoever. However subsequent investigations have shown that the universes in which all particle horizons are abolished form a set of measure zero amongst all possible universe models (MacCallum 1971) and further no mechanism for smoothening out high enough anisotropies seems available (Doroshkevich, Zel'dovich, and Novikov 1971).

More recently, it has been suggested (Zel'dovich 1972b) that the difficulty regarding this apparent breakdown of causality may be removed if, instead of a classical description, one changes over to a quantum theory (Misner 1973). As an example Zel'dovich points out that the particle creation process in a strong electric field appears to violate causality in a classical description, although the quantum theory of the process is completely consistent with causality. Further the hope has been expressed that particle creation in intense gravitational field of the early universe may serve to bring about isotropy in an initially anisotropic universe.

7
THERMAL HISTORY OF THE UNIVERSE AND NUCLEOSYNTHESIS

7.1. ELEMENTARY PARTICLE AND THE HADRONIC BIG-BANG

In conventional relativistic cosmology, the universe, near about the singular state, is considered to be constituted of radiation and ultrarelativistic elementary particles. The usual justification for such an idea arises from the fact that while the radiation density increases inversely as the fourth power of the linear dimensions, matter density increases inversely as the third power. Thus, unless radiation is altogether absent, one may expect the universe to be ultimately radiation dominated as the singularity is approached. The temperature would then become arbitrarily large and one has a 'hot big-bang'.

An investigation of the early universe, assuming it to consist of a limited spectrum of elementary particles like photons, gravitons, leptons, and quarks, was undertaken by Zel'dovich, Okun, and Pikel'ner (1966) and Novikov and Zel'dovich (1967). The hadrons were assumed to be composed of quarks and to be dissociated into quarks at high temperatures. At sufficiently high temperatures the quarks would be in thermal equilibrium with radiation, so that if n_{q}, n_{γ} is the number densities of quarks and photons at temperature

$$\frac{n_{\mathrm{q}}}{n_{\gamma}} \sim \exp(-m_{\mathrm{q}}/kT) \ .$$

Zel'dovich *et al*. obtained $n_{\mathrm{q}}n_{\gamma} \sim 10^{-18}$ at the present epoch and using the empirical value $n_{\gamma}/n_{\mathrm{B}} \sim 10^{9}$ (n_{B} is the baryon-number density, the value 10^{9} following from the background temperature 2·7 K for radiation and a plausible value of the matter density), one gets

$$\frac{n_{\mathrm{q}}}{n_{\mathrm{B}}} \sim 10^{-9}$$

which indicates that quarks should be as copious in the universe as gold nuclei. However, experimental limits on

n_q/n_B are much lower: $n_q/n_B < 10^{-19}$ (Stover, Moran and Trischka 1967; Chu, Kim, Beam, and Kwak 1970).

This apparent contradiction with the existence of quarks in hot big-bang cosmology is, however, resolved and one gets an altogether different picture of the early universe if one considers the big bang on the basis of ideas of modern elementary particle physics. Thus both the dual resonance and the 'bootstrap' models lead to a hadron level density formula for large mass values

$$N(m)\,dm \sim Am^{-B}\, e^{\beta m}\, dm$$

where A, B, and β are constants. The exponential factor in the level density formula means that as the energy of a hadronic system is increased, there would be an excitation of higher mass resonances rather than an increase of kinetic energy and the partition function as well as the total energy converges only if $kT < 1/\beta$. One thus gets an upper limit to the temperature which is $T_{\mathrm{lim}} \sim 2\times10^{12}$ K. (Hagedorn 1965, 1970; Lee, Leung, and Wang 1971; Frautschi, Steigman and Bahcall 1972; Tuan 1972).

As $T \to T_{\mathrm{lim}}$ the pressure tends towards a constant value while the energy density behaves as $(T_{\mathrm{lim}} - T)^{-\frac{1}{2}}$. Thus on the one hand in the hadronic big-bang, the temperature at the singularity is no longer infinite (warm big-bang) and the state of the matter is non-relativistic.

Under this circumstance, an integration of the Einstein equations is possible and Huang and Weinberg (1970) find that as $t \to 0$,

$$R \sim t^{2/3}(\ln t)^{1/3} ,$$

and the ultimate temperature is close to the limiting temperature:

$$\frac{T_{\mathrm{lim}}-T}{T_{\mathrm{lim}}} \to \frac{1}{\sigma} \sim 10^{-9}$$

where σ is the entropy per baryon and it is assumed that the universe does not have particle-antiparticle symmetry. Thus the large value of σ corresponds to the big-bang temperature being very nearly $T_{\lim}$.

The Huang-Weinberg calculation is based on the assumption that the baryonic chemical potential μ is much greater than $10^8\, kT_{\lim}$. This has been criticized as unrealistic by Stauffer (1972). If $1 \ll \mu(kT_{\lim})^{-1} \ll 1/\tau$ where $\tau \equiv (T_{\lim} - T)/T_{\lim}$, then Stauffer finds that for $\sigma \sim 10^9$, $\tau \geq 10^{-16}$. [See also Alexanian (1975); Alexanian and Mej'ia-Lira (1975)].

7.2. THE EARLY UNIVERSE

As the temperature drops below $\sim 10^{12}$ K, which according to Kundt (1971) occurs at time $t \sim 10^{-4}$ seconds, hadrons quickly disappear and the universe is dominated by a mixture of radiation and ultrarelativistic muons, electrons, and neutrinos of both types and their antiparticles. We may note two important points regarding this phase. Firstly, both for massless particles like the neutrinos and photons as well as for the muons and electrons, $p = \rho/3$ and the presure and density are so large that the cosmological terms as well as the curvature terms in equations (3.8) and (3.9) (p. 24) may be neglected. Thus we have quite generally the simple relations:

$$H \equiv \frac{\dot{R}}{R} = \left(\frac{8\pi\rho}{3}\right)^{\frac{1}{2}} \tag{7.1}$$

$$\rho R^4 = \text{const.} \tag{7.2}$$

$$RT = \text{const.} \tag{7.3}$$

Secondly, we shall show that the characteristic times for different processes bringing about thermal equilibrium are small compared to the time scale for the expansion of the universe which is H^{-1}. The scattering cross-section σ for electrons is of the order of the classical electron cross-section, i.e. $\sigma \sim 4\pi(e^2/mc^2)^2 \sim 10^{-24}$ cm^2. The characteristic time for the scattering processes involving electrons is thus $T = (n\sigma v)^{-1}$ where n is the number density of electrons and v is the relative velocity between the interacting particle and the electron. At the present epoch we may provisionally take the density of matter to be $\sim 10^{-30}$ g cm^{-3} and as this is predominantly hydrogen the number n of electrons or protons is $\sim 10^{-6}$ cm^{-3}. Now consider the phase for which $10^{12} \gtrsim T \gtrsim 5\times 10^{9}$ (the latter temperature being that for thermal production of electron positron pairs). As from equation (7.3) R/R_0 in this phase $\leq 10^{-9}$ (remember present temperature of radiation is 2·7 K) hence $n > n_0\, R_0^3/R^3 \approx 10^{21}$ cm^{-3}, we have given here the sign of inequality, for n will be significantly larger considering the electron-positron pairs that were present in this phase but have since been annihilated. The velocity v in this phase is effectively c, the velocity of light, so that $\tau < 10^{-7}$ seconds whereas H^{-1} being of the order of the age of the universe in this phase ranges from $\sim 10^{-4}$ to a few seconds. Thus $\tau \ll H^{-1}$ and hence the equilibrium which is brought about by scattering by electrons would persist throughout this phase.

One may wonder whether the equilibrium condition also obtains for neutrinos in view of the poor interaction of neutrinos with other particles. The following argument clarifies the situation in this respect (Kundt 1971).

Considering the neutrinos, electrons, and muons to be ideal gases their number densities will be given by the Fermi distribution formula;

$$n(p)\,\mathrm{d}p = 4\pi h^{-3} g p^2 \mathrm{d}p \left[\exp\left(\frac{E-\mu}{kT}\right) + 1\right]^{-1} ,$$

where $n(p)\,\mathrm{d}p$ is the number per unit volume in the momentum range p, $p+\mathrm{d}p$ and g is the spin statistical weight and is equal to 1 for neutrinos and 2 for electrons and muons, and μ is the chemical potential for the species. We shall assume for the moment that μ vanishes for all the species. For the ultrarelativistic region that we are considering, the energy $E \approx pc \gg m_0c^2$.

With $\mu = 0$, the number density falls off sharply for $E > kT$ and hence a particular species will be very poorly present at temperatures where kT is less than its rest mass energy m_0c^2. Thus for temperatures $T < 1\cdot 2\times 10^{11}$ K, the muons will be insufficient in number and the muonic neutrons will be decoupled from the photon-electron plasma. For the electronic neutrinos this decoupling would occur at $\sim 6 \times 10^9$ K. Above these temperatures, the number density of all the different species is given by

$$n \sim (KT/\hbar c)^3 ,$$

and for the cross-section for processes involving neutrinos and charged leptons we take (neutrino energy being large compared to the electronic rest mass energy, Bahcall 1964)

$$\sigma \approx f^2 \left(\frac{kT}{\hbar^2 c^2}\right)^2 \tag{7.6}$$

so that the frequency of these processes is of the order

$$Q = n\sigma\langle v\rangle = f^2 \hbar^{-7} (kT)^5 c^{-6} . \tag{7.7}$$

Here f is the weak coupling constant equal to $1{\cdot}02 \times 10^{-5}\, m_p^{-2}\, \hbar^3 c^{-1}$, m_p being the proton mass; so that $f = 1{\cdot}4 \times 10^{-49}$ cgs. Again the total energy density is

$$\rho = \rho_\gamma + 2\rho_\mu + 2\rho_e + 4\rho_\nu = \frac{K^4 T^4}{c^3 \hbar^3} . \tag{7.8}$$

Combining equations (7.1), (7.7), and (7.8) we get

$$\frac{Q}{H} \approx \hbar^{-11/2} f^2 (kT)^3 \approx \left(\frac{T}{10^{10}\,\mathrm{K}}\right)^3 . \tag{7.9}$$

Thus Q is large compared to H for $T > 10^{11}$ K and in these temperature ranges, the neutrinos would be in thermal equilibrium with the other constituents of the universe. Below this temperature the muonic neutrinos would first be decoupled and a little later the electronic neutrinos would follow. (See however de Graaf 1970; Weinberg 1967, 1972a; G't' Hooft 1971.)

Before pursuing the career of this photon-lepton combination further, we return for a moment to the chemical constant μ. The parameter μ arises in the distribution formula because of the constraint that the particle number must be conserved. For photons there is no such constraint and so μ does not appear, or in other words $\mu = 0$. For baryons and leptons, a pair may annihilate to produce photons - hence the chemical constants of a particle and its antiparticle must be equal in magnitude and opposite in sign. Looking back to equation (7.4) it is clear that if for any species μ be comparable with kT, then the net number density of the species would be comparable with the photon-number density at that temperature. As for the universe the baryon-photon ratio has the low value 10^{-9}, one may conclude that the chemical constant for baryons has a very small value and for simplicity can in most cases be made equal to zero. Similar arguments lead us to assume $\mu_e \approx 0$, but we have little precise knowledge regarding neutrinos. Of course, with the neutrino chemical constant significantly different from zero, one would expect a net excess of either neutrinos or antineutrinos. These would cause an impoverishment of the cosmic ray protons of high energy. From this consideration Cowsik, Pal, and Tandon (1964) concluded that the neutrino chemical constant μ_ν must lie between ± 2eV. Studying the proton spectrum data, Konstantinov and Kocharov (1964) and Konstantinov, Kocharov, and Starbunov (1968) gave $\mu_\nu \approx -0{\cdot}8$ eV. However such a value of μ_ν would lead to a high energy-density for neutrinos in the universe and make the present deceleration parameter $q_0 \approx 100$ (Weinberg 1972b). The currently popular low values of q_0 would almost force one to a vanishingly small value of μ_ν.

If, as seems likely, deuterons and helium nuclei were formed in the early stage of the big-bang universe, their abundances would depend sensitively on the initial neutron, proton ratio. This, in turn, would be affected if there was an excess of either neutrinos or antineutrinos. Thus, if one is to obtain reasonably good agreement with the observed abundances of deuteron and helium, one must have (Reeves 1972):

$$-0{\cdot}3 \leq L_e \leq +5$$

where

$$L_e \equiv \frac{(n_\nu - n_{\bar{\nu}}) + (n_{e-} - n_{e+})}{n_\gamma} \,. \tag{7.10}$$

If we take $n_{e-}-n_{e+} \approx n_B \approx n_\gamma\, 10^{-9}$, the above limits on L_e gives (with the present value of $n_\nu \approx 400$ cm^{-3}),

$$-10^{-4}\text{ eV} < \mu_\nu < 2\times10^{-4}\text{ eV},$$

limits which seem consistent with a low value of q_0.

With the μ's thus presumably vanishing, one can now with the help of equation (7.4), determine the number density, energy density, and entropy density as functions of the temperature T. So long as the temperature is high enough that the rest masses can be neglected in comparison with p/c, the values of these quantities are the same for all the fermions except for the weight factors g. Thus as

$$\int_0^\infty x^3(e^x+1)^{-1}dx = \frac{7}{8}\int_0^\infty x^3(e^x-1)^{-1}dx$$

the energy densities in the ultrarelativistic case are

$$\rho_\gamma = aT^4,\ \rho_{\nu_e} = \rho_{\nu_\mu} = \rho_{\bar{\nu}_e} = \rho_{\bar{\nu}_\mu} = \frac{7}{16}\,aT^4,\ \rho_{e+} = \rho_{e-} = \frac{7}{8}\,aT^4,$$

so that

$$\rho_e = \rho_\nu = \frac{7}{4}\,\rho_\gamma$$

In the regime where the muons are annihilated but electrons are still relativistic and the neutrinos have the same temperature as electrons and photons, we have for the entropies in volume V

$$\begin{aligned} S_\gamma &= \frac{4}{3}\,\rho_\gamma\,\frac{V}{T} = \frac{4}{3}\,aT^3V\ . \\ S_e &= \frac{4}{3}\,\rho_e\,\frac{V}{T} = \frac{7}{4}\,S_\gamma\ . \qquad (7.11) \\ S_\nu &= \frac{4}{3}\,\rho_\nu\,\frac{V}{T} = \frac{7}{4}\,S_\gamma\ . \end{aligned}$$

After the annihilation of electron-positron pairs, the neutrinos are decoupled and the temperature of radiation T_γ is no longer the same as the neutrino temperature T_ν. Again as the processes are reversible the total entropy remains constant but the entropy S_e goes to the radiation field increasing the radiation entropy by the factor 11/4, the neutrino entropy remaining unchanged. As $S \propto T^3$ for both neutrinos and photons, the ratio of their temperatures will be

$$\frac{T_\gamma}{T_\nu} = \left(\frac{11}{4}\right)^{1/3} = 1{\cdot}40 \ . \tag{7.12}$$

Put in another way the annihilations produce photons which enhance the radiation temperature and subsequently, due to expansion, both the neutrino and the photon temperature falls as R^{-1}; neutrino temperature continues to lag behind the radiation temperature maintaining the same ratio. With

$$\rho = \rho_\gamma + \rho_\nu + \rho_e = \frac{9}{2} aT^4 \ ,$$

equations (7.1) and (7.3) together give

$$\frac{\dot{T}}{T^3} = -(12\pi a)^{\frac{1}{2}}$$

which, on integration gives

$$T = \left(\frac{1}{48\pi a}\right)^{\frac{1}{4}} (t+t_0)^{-\frac{1}{2}} = \frac{10^{10}\ \mathrm{K}}{(t+t_0)^{\frac{1}{2}}}$$

In the radiation dominated elementary-particle model, the constant of integration t_0 is put equal to zero to signify that the temperature becomes infinite at the big bang. In our present discussion we shall construct the integration constant by the arbitrary choice $t = 0$ at $T = 10^{12}$ K which corresponds to the transition from the hadron dominated phase and occurs at $\sim 10^{-4}$ seconds after the big bang (Peebles 1966). As the temperature falls the rest-mass energy ceases to be negligible and one has to make numerical computations to evaluate T_γ and T_ν as functions of time. At this stage equations (7.2) and (7.3) are no longer valid but considering that the processes are all reversible, the total entropy $S_\gamma + S_e + S_\nu$ must remain constant throughout. This gives, with the help of equation (7.4), a functional relationship between T and V (or rather R). Another relation between T and R is provided by (7.1) with ρ again determined with the help of equation (2.4). Eliminating R between these two relations, one obtains a differential equation for T with time as the independent variable. This equation is numerically integrated, the constant value of entropy being fixed by the boundary condition $T_\gamma = 2{\cdot}7$ K and $T_\nu = 2{\cdot}7\times(4/11)^{1/3} = 1{\cdot}9$ K at $R = R_0$ (Peebles 1966).

At temperatures below 10^9 K, the electrons disappear and the photons and neutrinos undergo uncoupled but identical expansions. In this stage,

$$\rho = \rho_\gamma + \rho_\nu = aT_\gamma^4 + \frac{7}{4}\,aT_\nu^4 = 1{\cdot}5aT_\gamma^4$$

where we have used equation (7.12) T_γ now obeys the relation

$$T_\gamma = \left(\frac{1}{16\pi a}\right)^{\frac{1}{4}}(t+t_0')^{-\frac{1}{2}} = \frac{5{\cdot}10^9\ \text{K}}{(t+t_0')^{\frac{1}{2}}}\ .$$

This goes on until the energy density of matter becomes a significant part of the total energy density. So long as the matter is in the plasma state it is coupled with the radiation by the Thomson scattering process and the temperature of the matter and radiation may be taken as identical. However as the temperature falls, protons and electrons progressively combine to form hydrogen atoms. The equilibrium concentration α of ions is given by

$$\frac{\alpha^2}{(1-\alpha)}\,n = \left(\frac{m_e kT}{2\pi\hbar^2}\right)^{3/2} \exp\left(-\frac{13{\cdot}6\text{eV}}{kT}\right) = 2{\cdot}4\times10^{15}T^{3/2}\exp\left(-\frac{1{\cdot}6\times10^5}{T}\right)$$

where

$$\alpha = \frac{n_p}{n} = \frac{n_p}{n_H + n_p}$$

n_p and n_H being the number densities of protons and hydrogen atoms respectively and it is assumed that the plasma is neutral $n_p = n_e$. For $\alpha = \frac{1}{2}$, Table 7.1 gives the values of temperature T for various values of n (equivalently of the matter density ρ).

TABLE 7.1

Temperature for half ionization of hydrogen at different densities

Density ρ ($g\ cm^{-1}$)	10^{-24}	10^{-16}	10^{-8}	10^{-4}	10^{-2}
$n = 2n_p$	0·6	6×10^{7}	6×10^{15}	6×10^{19}	6×10^{21}
T	$3{\cdot}2\times10^{3}$	5×10^{3}	11×10^{3}	27×10^{3}	61×10^{3}

It is interesting to note that half ionization occurs not for $KT \approx 13{\cdot}6$ eV but at a much lower temperature due to the large coefficient of the exponential term. Also T for half ionization is not very sensitive to variation of n (Zel'dovich and Novikov 1971). As a convenient working value we take 4000 K to be the temperature of recombination of protons and electrons after which the matter (decoupled from radiation) may be considered to be a monatomic gas obeying the relation $T_{\text{mat}} R^2$ = constant whereas the radiation temperature follows $T_\gamma R$ = constant. Thus the matter temperature falls faster and at say $R = R_0/3$ ($z = 2$), $T_{\text{mat}} \ll T_\gamma = 8{\cdot}1$ K. However in our study of the search for missing matter in Chapter 4, we have seen that if the density of hydrogen is not to be too low, it must be ionized at $z = 2$. This would mean that at some stage, the steady fall of temperature of the gas has not only been arrested but actually a reheating has occurred. We shall presently examine this problem of reheating but before that we give the results of the computations by Peebles regarding T_γ and T_ν.

Table 7.2 shows the typical expansion of a mixture of radiation and ultrarelativistic particles up to $0{\cdot}6 \times 10^{10}$ K ($RT_\gamma = 1{\cdot}9\ R_0$). After that up to $0{\cdot}2 \times 10^{10}$ K, the electrons have a pressure p_e less than $\rho_e/3$, $\gamma \equiv C_p/C_v$ for the electron gas is greater than 1·33 and RT_γ shows a progressive decrease.

TABLE 7.2

Variation of the radiation temperature, neutrino temperature, linear dimensions with time.

Time (s)	T_γ (10^{10} K)	T_γ/T_ν	R/R_0	$T_\gamma R/R_0$ (K)
0	100	1·000	$1{\cdot}9\times10^{-12}$	1·9
$1{\cdot}94\times10^{-4}$	60	1·000	$3{\cdot}2\times10^{-12}$	1·9
$1{\cdot}13\times10^{-3}$	30	1·000	$6{\cdot}4\times10^{-12}$	1·9
$2{\cdot}61\times10^{-3}$	20	1·000	$9{\cdot}6\times10^{-12}$	1·9
$1{\cdot}08\times10^{-2}$	10	1·000	$1{\cdot}9\times10^{-11}$	1·9
$3{\cdot}01\times10^{-2}$	6	1·000	$3{\cdot}2\times10^{-11}$	1·9
0·121	3	1·001	$6{\cdot}4\times10^{-11}$	1·9
0·273	2	1·002	$9{\cdot}6\times10^{-11}$	1·9
1·103	1	1·008	$1{\cdot}9\times10^{-10}$	1·9
3·14	0·6	1·022	3.1×10^{-10}	1·9
13·83	0·3	1·081	$5{\cdot}9\times10^{-10}$	1·8
35·2	0·2	1·159	$8{\cdot}3\times10^{-10}$	1·7
182·0	0·1	1·346	$2{\cdot}6\times10^{-9}$	2·6
$2{\cdot}08\times10^{3}$	0·03	1·401	$9{\cdot}0\times10^{-9}$	2·7
....	$2{\cdot}7\times10^{-10}$	1·401	1	2·7

From 0·2 to 0·03 10^{10} K the annihilation of electron-positron pairs heats up the radiation so that RT_γ increases to its present value 2·7. The decoupling of the neutrinos from the photons begins at about 3×10^{10} K and is complete at about 3×10^{8} K. After this T_γ/T_ν maintains the constant value 1·40.

7.3. THE REHEATING OF THE COSMIC GAS

We return to the question of reheating. It is well to emphasize that reheating is necessary for ionization of hydrogen and that again is demanded only if we like to have a high density of hydrogen for closure of the universe. But with the present trend towards a low value of q_0 and a low density universe one may very well be reconciled with an absence of hydrogen in the intergalactic space and consider the need for reheating redundant. Keeping this in mind, let us review the arguments regarding reheating.

Ginzburg and Ozernoy (1966) as well as Weymann (1967) have contemplated that the reheating is brought about by cosmic rays from active galaxies. As one requires the high degree of ionization at $z = 2$ (to explain the observational data on 3C9) Weymann considers the heating to take place at somewhat larger values of z. Field (1969) estimates that to heat the gas to temperatures above 10^6 K about 10^{15} ergs per gram of the gas is necessary. Much of this would be radiated away mostly in the form of Lyman α-lines of H and He^+ and may be expected to be present in the form of radiation in the wavelength region 10^3 - 10^4 Å due to red shift. Again in the Waymann model, as the mass of intergalactic gas is about fifty times the mass of visible matter in the galaxies, the energy requirement works out to 6×10^{16} erg g^{-1} for the galaxies, or about 10^{61} erg per galaxy. Thus the average galaxy must produce about this much energy in the form of cosmic rays. As this is too high a figure relative to the present cosmic-ray energy production in our galaxy, the Weymann model can be correct only if most galaxies have gone through an explosive phase (like the radio galaxies) sometime in their past.

Another point of interest is the stage at which the heating might have taken place. The departure (or rather the lack of departure) of the microwave background from the Planck distribution would provide a useful clue on this point. If the electron, set free as a result of the heating, is at a significantly higher temperature than the radiation, then the inverse Compton process leads to a redistribution of the photon energies such that the number of high frequency photons as well as the total radiation energy density is increased above the values determined by the black-body radiation formula with T_γ calculated from the observed intensity in the low frequency region (Sunyaev and Zel'dovich 1969).

Both these effects will be enhanced as the number of scattering electrons, or in other words the density of matter, increases at the stage of ionization. In this way Weymann (1966) concluded that for the Einstein - de Sitter universe the heating and re-ionization might have occurred only later than $z = 200$, if one is not to run into contradiction with observations.

7.4. NUCLEOSYNTHESIS AND THE ABUNDANCE OF LIGHT ELEMENTS

We shall first give an outline of the theory of formation of light nuclei in the early stage of the big-bang universe and then review the observational data. The theory of synthesis of the different nuclear species in the interior of stars as developed by Burbidge, Burbidge, and Fowler (1957) (see also Hoyle 1954) was remarkably successful in explaining the abundances of heavier nuclei, but the theory gave too poor abundances for lighter nuclei. Some of these lighter nuclei such as ^{6}Li, ^{9}Be, ^{10}B, and ^{11}B could be produced by cosmic rays and one could thereby account for their observed abundances (Reeves, Fowler, and Hoyle 1970; Meneguzzi, Audouze, and Reeves 1971). However one cannot explain in this way the observed abundances of ^{2}H, ^{3}He, ^{4}He, and ^{7}Li (Reeves, Audouze, Fowler, and Schramm 1973).

In the late forties, Gamow and his school proposed a theory of nucleogenesis in which the early universe near the big bang was suggested as the locale (Alpher, Bethe, and Gamow 1948; Alpher and Herman 1949; also unpublished works of Fermi and Turkevich). With an initial pure neutron gas, successive nuclei were thought to be built up by neutron capture and the charges were subsequently adjusted by β-decay. The reactions stopped as the neutrons were impoverished by decay and capture. The theory faced serious difficulties in crossing the mass numbers 5 and 8 and thus could not give appreciable abundances for heavier nuclei. With the success of the stellar production theories, the α-β-γ theory (as Gamow's theory was called) was all but forgotten.

However, the difficulty regarding the light nuclei that one faces in stellar theories, led to a re-investigation and refined version of the α-β-γ theory by Peebles (1966) and a little later by Wagoner, Fowler, and Hoyle (1967). Since then

Hayashi (1950) has pointed out that at temperatures $T \gtrsim 10^{10}$K, the neutrons and protons are rapidly brought into thermal equilibrium through the weak interaction reactions

$$p+\bar{\nu} \rightleftarrows n+e^{+}, \quad p+e^{-} \rightleftarrows n+\nu, \quad n \rightleftarrows p+e^{-}+\nu. \tag{7.13}$$

If none of thc leptons are degenerate (i.e. $\mu/KT \ll 1$) so that the particle-antiparticle concentrations are equal, one would have at temperatures $T \gtrsim 10^{10}$ K, the neutron-proton ratio given by their equilibrium value

$$\frac{X_{\mathrm{p}}}{X_{\mathrm{n}}} = \exp[(m_{\mathrm{n}}-m_{\mathrm{p}})c^{2}/KT]$$

where X_{p}, X_{n} are the numbers of protons and neutrons respectively at temperature T. With $m_{\mathrm{n}}-m_{\mathrm{p}} = 1{\cdot}3$ MeV, this gives nearly 38 for the percentage concentration of neutrons. In general, when the temperature is not so high, one may calculate the concentration of neutrons as a function of time (or equivalently of temperature) knowing the reaction rates of the $n \rightleftarrows p$ processes. Such a calculation was indeed done by Peebles (1966) disregarding the formation of any other nuclei.

Wagoner *et al.* (1967) started their calculation at $T = 6{\times}10^{10}$ K and the reactions proceed up to $T = 10^{8}$ K. (In a later paper Wagoner (1973) took the temperature up to $T = 10^{11}$ K.) At these initial temperatures, all the thermonuclear reactions are fast enough to allow one to assume that the different nuclear species are in thermal equilibrium. Consequently their abundances are governed by Boltzmann factor leading to a practically negligible concentration of all nuclei other than protons and neutrons. The equations governing the nucleosynthesis may be schematically represented in the following manner:

Rate of change of the ith nuclear species	=	Rate of change due to universal expansion	+	Rate of change due to the species being used up in reactions leading to other species	+	Rate of change of the species due to its production by reaction between others. (7.14)

The first two factors on the right lead to a diminution and the third to an increase in the number density of the ith species. Of course the last two terms would lead to a coupling between the equations governing the different species. The first factor depends directly on the expansion rate $\dot{R}/R$ and we have already seen how it can be calculated for the early stages. For the other two factors one needs the concentrations of different species and the cross sections of the reactions involved. Wagoner *et al.* (1967) considered a total of 144 reactions involving nuclei up to mass number 23. The cross sections are functions of temperature and thus in general decrease with time due to expansion. The concentrations of nuclei will of course depend, besides all these factors on the amount of matter or rather the total baryon number in the universe.

As the discussion in the previous section has shown, it is possible to calculate the temperature as a function of time. Combining the known functional relationships between R, T, and time t, one can integrate the system of equations (4.3), the only free parameter being the baryon mass density ρ_b. However we have

$$\rho_b \propto n_b \propto n_\gamma \propto T_\gamma^3$$

so that we can write $\rho_b = hT_\gamma^3$ and with the present value of $T_\gamma = 2{\cdot}7$ K and plausible values of matter density, h would lie between $10^{-32} - 10^{-30}$ g cm^{-3} (K)$^{-3}$. The calculation of Wagoner *et al.* showed that the resultant abundances of ^{4}He are not very sensitive to the value of h. Thus for

$10^{-32} \lesssim h \lesssim 10^{-27}$, the ^4He abundance has a value near 27 per cent by mass. However ^3He and ^2H abundances are extremely sensitive - while for $h \approx 10^{-32}$, the ^2H abundance has nearly the value $\approx 10^{-4}$ times the abundance of ^1H, the ^2H abundance falls off sharply as h increases.

Recently Wagoner (1973) has 'revisited' the problem with a revised value of the neutrino half life (Christensen, Nielsen, Bahnsen, Brown, and Rustad 1967) and some other improvements and with 'a greater confidence that the microwave background is indeed a relic of the big bang'. He finds that the possible abundances of all the three (^2H, ^3He, and ^4He) fit in rather nicely with the observational data if the present matter density is $(1 - 3) \times 10^{-31}$ g cm^{-3}. We may easily understand the necessity for a low baryon density in order to account for the ^2H abundance in the following manner. The deuteron nuclei are formed by the reaction

$$n + p \rightarrow {}^2H + \gamma$$

and are consumed in a number of helium producing reactions

$$^2H + {}^2H \rightarrow {}^3He + n,$$

$$^2H + {}^2H \rightarrow {}^3H + p,$$

$$^3H + {}^2H \rightarrow {}^4He + n \ .$$

All these deuteron consuming reactions have large cross-sections so that an increase of baryon-number density allows them to proceed rapidly resulting in a rapid depletion of ^2H nuclei. Indeed as Wagoner (1973) shows, one has the relation

$$(\rho_b)_0 \cdot X\,(^2H) \leq 2\times10^{-34} \text{ g cm}^{-3}$$

where $(\rho_b)_0$ is the present matter-density (at $T_\gamma = 2{\cdot}7$ K) and $X(^2H)$ is the deuteron abundance measured as the ratio $N(^2H)/N(^1H)$. Thus to have $X(^2H) \approx 2\times10^{-4}$ (the desired value) one must have $(\rho_b)_0 < 10^{-30}$g cm^{-3}.

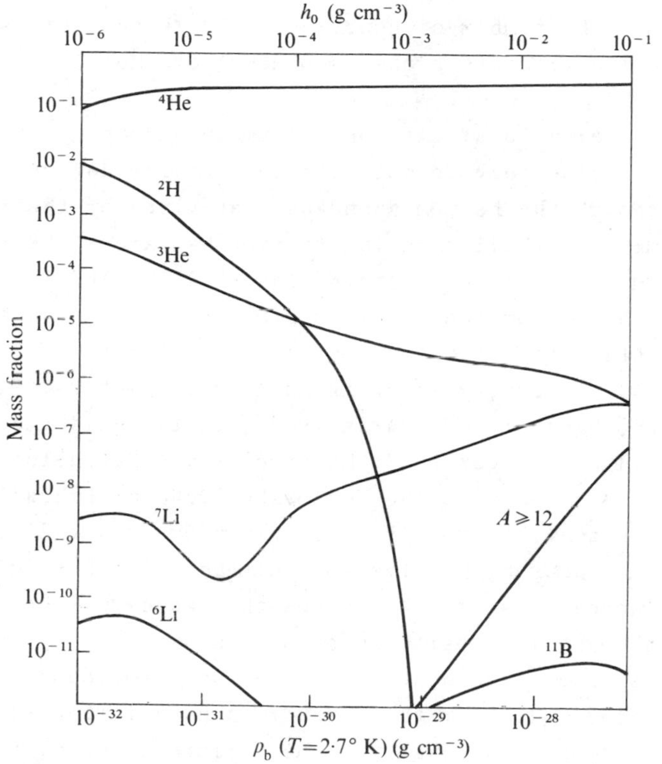

Fig.7.1. Abundances of light elements for different values of the present baryonic density ρ_b showing the particular sensitiveness of the deuteron abundance. (Wagoner, 1973.)

We may note in passing that the very large cross-section of the deuteron consuming reactions is also the reason why the ^{2}H abundance is reduced to a poor value in the stellar theory of nucleogenesis.

It is also fairly easy to understand the insensitivity of the helium abundance to the matter density. The nucleo-synthesis can begin only if the deuterons do not undergo fast enough photodisintegration. Thus nucleosynthesis begins rather sharply at $T = 10^9$ K. Now, provided the baryon density is large enough for the dueteron - deuteron reactions to occur, practically all the neutrons present at 10^9 K go to form ^{4}He. Thus the number of ^{4}He nuclei formed should be nearly half the number of neutrons at 10^9 K and this simple consideration leads to a mass abundance of ^{4}He near 27 per cent.

The observational data on helium abundance at various locales unfortunately do not give very concordant results. Thus, although the helium abundances as given by the ratio of the number of helium nuclei to protons, are ~0·08 to 0·10 in the sun, ~0·10 in most stars, and ~0·11 ± 0·03 in the interstellar medium and are also similar in galactic and extragalactic HII regions and planetary nebulae, the abundance is significantly lower in many halo Bp stars (Greenstein 1966; Sargent and Searle 1966), in the galactic centre (Churchwell and Mezger 1973; Huchtmeier and Batchelor 1973), and in quasars (Bahcall and Kozlovsky 1969a,b; Bahcall and Oke 1971; a contradictory view is given by Jura 1973). In the case of the sun, observations of the solar cosmic rays and prominences lead to the values that we have mentioned above; the mass luminosity relationship as given by the usual theories of energy generation in the sun also leads to a similar value. However these theories have lately become subject to some doubt owing to the failure to observe the expected neutrino flux from the sun (Davis *et al.* 1968; Davis 1972; Kuchowicz 1976c).

In a recent survey of 39 bright galactic HII regions in the northern and southern hemispheres, Churchwell, Mezger, and Huchtmeier (1974) studied the radio recombination lines of hydrogen and helium. The observed regions were up to a distance of 13 Kpc from the galactic centre. It was concluded that the abundance of helium in the galaxy is approximately constant and comparison with abundances in other objects did not reveal any significant difference. The currently observed helium abundance was considered a superposed effect - a part being produced in the pregalactic stage by processes we have just described and another part produced in stars. The primeval abundance of ^{4}He was suggested to be 0·08. The investigations did not clarify the point as to whether the low ^{4}He abundance in the galactic centre obtains only in the central region or there is a progressive decrease in the ^{4}He abundance as one approaches the centre.

For an argument in favour of the production of helium in the pregalactic stage rather than in the stars, we may cite the observation of Searle and Sargent (1972). They found that the abundance in two dwarf blue galaxies, which are apparently young systems, of helium is 27 to 31 per cent while the abundances of heavy elements are significantly lower than in the sun.

Regarding the abundance of ^{3}He, in the galactic HII regions the ratio $N(^3\mathrm{He})/N(\mathrm{H}) < 5\times10^{-5}$ (Seling and Heiles 1969; Predmore, Goldwire, and Walters 1971) while for the sun the ^{3}He abundance has been estimated as $2 \pm 1\times10^{-5}$ (Reeves *et al.* 1973; Cameron 1973, quoted by Talbott and Arnett 1973).

In a way the data regarding deuteron abundance are more significant than those for helium abundance because, in general, processes in stellar evolution would lead to a destruction of ^{2}H rather than to its production and according to the estimate of Thuan and Ostriker, the destruction may range from 10 to 90 per cent. Black and Dalgarno (1973) give for the zeta Oph cloud $2\times10^{-6} < N(^2\mathrm{H})/N(\mathrm{H}) < 2\times10^{-4}$ while Cesarsky, Moffet, and Pasachoff (1973) put the same limits to be 3×10^{-5} and 5×10^{-4} for the galactic centre region and Jefferts, Penzias, and Wilson (1973) and Wilson Penzias, Jefferts, and Solomon (1973) give $N(^2\mathrm{H})/N(\mathrm{H}) \sim 3\times10^{-3}$ for the Orion nebula. Again from an observation of absorption lines in the spectra of beta Centauri, Rogerson and York (1973) find $N(^2\mathrm{H})/N(\mathrm{H}) \sim 1{\cdot}4 \pm 0{\cdot}2\times10^{-5}$, while Spitzer, Drake, Jenkins, Morton, Rogerson, and York (1973) give $N(^2\mathrm{H})/N(\mathrm{H}) \sim 5\times10^{-3}$. These 'high' values of deuteron abundance, close to that one observes on the earth and in the sun (Cameron 1973; Reeves *et al.* 1973) may be explained by the big-bang nucleogenesis theory, as we have seen, provided we assume a low value for the present density of matter in the universe. While this is consistent with the present evidence in favour of a low q_0, there is a body of opinion which find it difficult to accept that there is a virtual absence of any matter in the intergalactic space. Thus alternative locales for deuteron production have been occasionally suggested (Fowler and Hoyle 1973; Colgate 1973) but such theories are apparently not quite satisfactory. In a recent note Ostriker and Tinsley (1975) have proposed an observational criterion to decide between the stellar and cosmological origin of deuteron - there should be a positive correlation between deuteron abundance and metal abundance in different regions of the galaxy if the origin is stellar, whereas there should be an anti-correlation if the origin is near the big bang.

It is important to look back to the assumptions that have been made in the theory of cosmological nucleogenesis. We have assumed the isotropic homogeneous model to calculate the expansion and the temperature history. Secondly we have assumed that none of the leptons are degenerate and thirdly we have considered the neutrinos as decoupled from photons and electrons for $T \lesssim 10^{10}$ K.

The influence of anisotropy, as distinct from non-homogeneity, has been investigated by Thorne (1967) and somewhat earlier by Hawking and Taylor (1966). The presence of shear increases the deceleration (cf. the Raychaudhuri equation) and thus affects the abundances. Thorne examined the Bianchi type I models - the magnitude of anisotropy in the early stages being limited by the consideration that the anisotropy of the radiation background must not exceed the limits set by observation. An the time of Thorne's investigation this limit was rather modest - about 3 per cent but as we have seen earlier, recent observations have brought it down to about 0·1 per cent. Thorne found that for an appreciable anisotropy lasting for a comparatively short time (~100 years) the ^{4}He abundance is greatly increased. Indeed for an anisotropy lasting for a period in between 0·1 and 10 years, the universe would have contained practically only ^{4}He. As the time during which an appreciable anisotropy persists increases, the helium abundance decreases and there would be hardly any helium if the anisotropy remains for 10^5 years or more. (In a more recent calculation, Barrow (1976) claims that the constraints of the deuteron and ^{4}He abundances require that the present ratio of shear to expansion $(\sigma/\theta)_0 \lesssim 4{\cdot}8 \times 10^{-12}$.)

Although the Thorne discussion was limited to homogeneous universes of a particular type, it is reasonable to suppose that the introduction of a shear in the early stages would seriously affect the calculated abundances and one may, by suitably choosing the parameters, obtain both higher and lower ^{4}He and ^{2}H abundances.

We have already noted that Reeves (1972) was able to set some limits on L_e (for definition see equation 7.10) using the idea of cosmological origin of deuteron and helium and their observed abundances. However in a recent paper the same problem of lepton degeneracy has been considered from a different angle by Yahil and Beaudet (1976). They have shown that with a non-zero lepton number one can obtain agreement with the observed ^{4}He and ^{2}H abundances with much higher values of the matter density in the universe. Thus, according to them, it is not correct to say that the observed deuteron abundance forces one to a low-density universe.

It is fairly easy to see qualitatively how a neutrino (or antineutrino) degeneracy would affect nucleosynthesis. The reactions (7.13) which bring about thermal equilibrium between neutrons and protons cannot proceed unimpaired in either direction if there be an excess of neutrinos (or antineutrinos). Suppose for example the antineutrinos are degenerate. Then the disintegration of neutrons

$$n \rightarrow p+e^{-}+\bar{\nu}$$

can occur only if the product antineutrino has an energy above the Fermi energy for the degenerate antineutrinos and for a similar reason the reaction

$$p+\bar{\nu} \rightleftarrows n+e^{+}$$

can also proceed only in the forward direction. Thus there would be a population of neutrons alone and so the deuterons cannot be formed. As the expansion dilutes the antineutrinos, the neutrons would begin to decay, but the temperature would then be too low for thermonuclear reactions and unless the density still remains high enough for appreciable deuteron formation we should have a pure hydrogen universe.

In the opposite case of neutrino degeneracy, there would be a preponderance of protons and again we should have a hydrogen universe. Thus apparently the low abundances of helium in some locales may be an indication of a neutrino (or antineutrino) degeneracy.

The influence of the 'decoupling' between electrons and neutrinos, as assumed in the standard calculation, has been examined by Hecht (1971, 1973). He finds that even if one assumes the coupling to persist throughout the nucleogenesis (due presumably to a large electron-neutrino scattering cross section), the helium abundance can at most be marginally affected.

Silk and Shapiro (1971) have shown that quite large variations in helium abundance may be explained by considering temperature fluctuations in the early universe. However a contradictory result has been obtained by Gisler, Harrison, and Rees (1974). (For a study of the influence of simultaneous presence of baryons and antibaryons on helium abundance see Schatzmann (1970); see also Epstein and Petrosian (1975).)

8
THE SINGULARITY OF COSMOLOGICAL MODELS

8.1. THE EXISTENCE OF COSMOLOGICAL SINGULARITY

The Friedman models, with or without pressure, have a singularity in the finite past (or in the finite future for time reversed contracting universes) if $\Lambda \leq 0$. This follows from the relation

$$\frac{\ddot{R}}{R} = \Lambda - \frac{4\pi}{3}(\rho + 3p) \leq 0 \; .$$

Thus R has no minimum and a decreasing R goes to zero in a finite time bringing about an infinity in the density as well as the curvature. To Einstein (1950) this appeared to be a sympton of the basic limitation of the general theory of relativity.

> The theory is based on a separation of the concepts of the gravitational field and matter. While this may be a valid approximation for weak fields, it may presumably be quite inadequate for very high densities of matter. One may not therefore assume the validity of the equations for very high densities and it is just possible that in a unified theory there would be no such singularity.

However a unified theory of the type Einstein conceived has not been constructed even now and by and large the overwhelming majority of physicists entertain serious doubts whether a unified-field theory of the pattern Einstein worked on in his latter life is at all realizable (Oppenheimer 1971; Pauli 1958; Schwinger 1976).

At first sight, the responsibility for the singularity may be laid at the door of the assumptions of homogeneity and isotropy (Eddington 1939; Tolman 1949). However it was soon discovered that, irrespective of homogeneity and isotropy, any fluid distribution whose motion is geodetic and irrotational would have a singular state of infinite density in its career (Raychaudhuri 1955a). This follows directly from the Raychaudhuri equation (equation 5.6, p. 81). With the acceleration $\dot{v}^{\alpha}$ and vorticity ω^{α} vanishing, the shear only augments the gravitational action of p and ρ and thus brings about a more rapid collapse to the singular state of infinite density.

To investigate the question of singularity in more general cases, one needs a definition of singular space-time (Geroch 1970a; Hawking and Ellis 1973). Simply singular behaviour of metric tensor-components are of course of little importance: e.g. a spherically symmetric line-element can be written in the form

$$ds^2 = g_{00}dt^2 - g_{11}dr^2 - r^2(d\theta^2+\sin^2\theta d\phi^2) .$$

Here g_{22} and g_{33} vanish at $r = 0$ and the corresponding contravariant g^{22}, g^{33} go to infinity. However one can in many cases of spherical symmetry get a perfectly regular behaviour at $r = 0$ simply by making a transformation. One may think that a satisfactory criterion would be to examine the behaviours of scalars formed from the Riemann-Christoffel tensor, e.g. the curvature scalar $R_{\alpha\beta\sigma\mu}R^{\alpha\beta\sigma\mu}$. Surely an infinity in the value of a scalar cannot be smoothed out by a transformation. However one may just get rid of the region where the scalar shows singular behaviour and the truncated space may camouflage as the complete space. Consider for example the Schwarzshild metric

$$ds^2 = \left(1 - \frac{2m}{r}\right)dt^2 - \left(1 - \frac{2m}{r}\right)^{-1} dr^2 - r^2(d\theta^2+\sin^2\theta d\phi^2) .$$

The so called 'Schwarzchild singularity' at $r = 2m$ is only a co-ordinate singularity but the field has a singularity of $R_{\alpha\beta\sigma\mu}R^{\alpha\beta\sigma\mu}$ at $r = 0$. Now introduce a new radial variable by the relation

$$r - 2m = r^{*2} .$$

The Schwarzschild metric assumes the form

$$ds^2 = \left(\frac{r^{*2}}{r^{*2}+2m}\right)dt^2 - (r^{*2}+2m)(d\theta^2+\sin^2\theta d\phi^2) - 4(r^{*2}+2m)dr^{*2} .$$

In the truncated space $0 \le r^* \le \infty$, there would obviously be no singularity of $R_{\alpha\beta\sigma\mu}R^{\alpha\beta\sigma\mu}$. In fact in the familar case of the transformation to the 'isotropic co-ordinate'

$$r = (1+m/2\bar{r})^2 . \bar{r}$$

the Schwarzschild line-element assumes the form

$$ds^2 = \frac{(1-m/2\bar{r})^2}{(1+m/2\bar{r})^2}dt^2 - (1+m/2\bar{r})^4(d\bar{r}^2+\bar{r}^2d\theta^2 + \bar{r}^2\sin^2\theta\, d\phi^2)$$

and in the new co-ordinate system the region containing the singularity $0 \le r \le 2m$ has been simply excluded and the region $2m \le r \le \infty$ has been mapped-out twice, once in the region $\frac{1}{2}m \le \bar{r} \le \infty$ and again in the region $0 \le \bar{r} \le m/2$.

It is thus not always correct to regard the absence of infinities in the value of scalars as a sure criterion of the regularity of space time. One defines a space time to be non-singular only if all time-like and null geodesics are complete in the sense that they can be extended to arbitrary values of their affine parameter. A time-like (or null) geodesic gives the path of a free massive or massless particle. Thus the definition may be taken to mean that the universe must contain the complete history of such particles. Put in another way, there must not be either a beginning or an end of time for any free test-particle. However the definition is silent about non-geodetic time-like lines and Geroch (1967) has indeed provided an example where the space time, though geodesically complete, nevertheless contains time-like lines with bounded acceleration which are of finite length. (For a discussion of space-like geodesic completeness, see Schmidt 1973).

A fairly large amount of literature exists on singularity theorems in cosmological models, but in most of the older discussion there is some global assumption like homogeneity or the existence of a Cauchy surface. These assumptions, by their very nature, are not subject to direct observational verification. However the microwave background-radiation provides us with valuable knowledge which has enabled Hawking and Ellis (1968) and more recently Hawking and Penrose (1970) to conclude that the actual universe must have singularities without using any global assumption except that of causality. The Hawking - Penrose theorem (for an exhaustive and authoritative discussion of the singularity theorems see Hawking and Ellis 1973) states that space-time singularities occur if there is a point p whose past null-cone encounters sufficient matter that the divergence of the null rays through p changes somewhere to the past of p (i.e. there is a minimum apparent solid-angle viewed from p for small objects of a given size). The theorem is true provided, (i) the Einstein equations hold with $\Lambda \leq 0$, (ii) the energy condition $T_{\mu\nu} v^\mu v^\nu \geq T/2$ holds for all time-like unit vector v^μ - this means in particular that for a perfect fluid $\rho+p \geq 0$ and $\rho+3p \geq 0$ (this follows because the scalar product of two time-like unit vectors v^μ, u^μ satisfy $v^\mu u_\mu \geq 1$ and may be arbitrarily large), (iii) there are no closed time like lines which would mean a violation of causality, and (iv) on each time like or null geodesic there is at least one point for which

$$K_{[a}R_{b]cd[e}K_{f]}K^c K^d \neq 0$$

where K_a is the tangent to the curve at the point under consideration - this condition apparently holds true for any general solution. Now Hawking and Penrose assume from the observed Planck distribution of energy in the background radiation that there must be sufficient matter on each past directed null geodesic from p (i.e. our present position in space time) to make the optical depth large so that the radiation undergoes repeated scattering before reaching us. This assumption leads to the fulfilment of the condition regarding the change of sign of the divergence of the null congruence through p, and thus one is led to conclude the existence of space-time singularities.

Hawking and Ellis, on the other hand, assume on the basis of the isotropy of the background radiation that the universe is of the Friedmann type up to the last scattering epoch and are thereby able to prove the existence of singularities.

The Hawking-Penrose theorem is, however, silent about the nature of the space-time singularity apart from geodetic incompleteness. It may well be that the singularity consisted of isolated points or time-like surfaces which most of the world lines manage to avoid. (Such a situation occurs in the Kerr metric where only few geodesics hit the singularity.) Indeed Hawking and Ellis speculate that prior to the present expansion, there was a collapsing phase. In that phase local inhomogeneities grew large and isolated singularities occurred. Most of the matter avoided the singularities and re-expanded to give the presently observed universe.

An example has been given by Shepley (1969) where a universe admitting the Bianchi type V group of motions has a singularity without the matter density going to infinite values. Shepley's line element

$$ds^2 = 2dtdx + a\,dx^2 + b^2e^{2x}(dy^2+dz^2)$$

where a and b are functions of t alone, was previously given by Farnsworth (1967). More recently Ellis and King (1974) and King (1974) have shown that some spatially homogeneous universes have a singularity where all physical quantities remain well behaved. For universes in which there is neither acceleration nor rotation as also for Bianchi type IX universes, the singularity is, however, physical (Matzner 1970; Matzner, Shepley, and Warren 1970). The Shepley-Farnsworth universe is a dust universe without rotation and hence besides the singularity studied by Shepley, there must be a physical singularity of infinite density as well. (Raychaudhuri, 1955a.)

The 'whimper' singularity of Shepley and of Ellis and King are associated with a situation in which the time-like geodesics normal to the homogeneous three-spaces all simultaneously approach the null direction and then these normal geodesics become space like, i.e. using a synchronous metric $ds^2 = g_{00}dt^2+g_{ik}dx^idx^k$, a whimper singularity is associated with a sudden change in g_{00} from +1 to -1 (Evans 1975).

Practically the only ways to escape from the Hawking-Penrose singularity theorem within the frame work of the general theory of relativity seem either to accept closed time-like lines implying a breakdown of causality (cf. the Gödel universe) or to consider departures from the energy condition. Cosmological solutions without causality can hardly be accepted as models of the actual universe. Departures from the energy condition may be in the form of negative energy as the C field of Hoyle and Narlikar (1970) or the negative pressure of McCrea (1951). The singularity free models, presented by Murphy (1973), Heller *et al.* (1973), and Heller and Suszycki (1974) all have bulk viscosity in a form so as to violate the energy condition. A particle creation in the gravitational field may also give rise to negative pressure terms leading to a violation of Hawking-Penrose condition (Parker and Fulling 1973).

A novel way of bringing in effectively negative energy has been pointed out by Marochnik, Pelikhov, and Vereshkov (1975). Their idea is to consider turbulence in the early universe. The turbulence causes fluctuation both in the metric and the fluid variables. They then adopt an averaging procedure for the Einstein field equations. This brings in some additional terms which under certain conditions simulate negative energy and may arrest the collapse to the singularity. (see also Nariai 1974a,b, 1975a,b).

The above approaches towards non-singular models do not find general acceptance and suspicion has frequently been expressed that the root of the trouble is in the disregard of quantum effects at large concentrations and small distances. Thus quantum effects are apparently important for lengths $l \sim (G\hbar/c^3)^{\frac{1}{2}} = 1{\cdot}6\times10^{-33}$ cm and the corresponding time is $t \sim l/c = 0{\cdot}5\times10^{-43}$ s. The density at this epoch would be $\rho \sim c^5/\hbar G^2 = 5\times10^{93}$ g cm^{-3} (Ginzburg 1971). Bahcall and Frautschi (1971) point out that if the universe in its early stages was hadron dominated then $p/\rho \to 0$ and the expansion rate $\dot{R}/R \approx 2t^{-1}/3$ so that if the velocities across the surface of the hadron (linear dimension $\sim10^{-13}$ cm) is not to exceed c, $t > 10^{-23}$ s. Thus it is reasonable to expect that the singularity will be profoundly affected and may even be prevented by the properties of the hadrons. On the basis of some simplified quantum calculations Misner (1969a) was led to believe that quantum effects would not change the occurrence or nature of the cosmological singularity. However we must emphasize that a quantized theory of gravitation still remains unrealized.

In Chapter 9 we shall consider the position of singularity in the Einstein-Cartan theory and the impact of the two tensor theory of Isham and collaborators.

8.2. THE OSCILLATORY APPROACH TO SINGULARITY IN SOME ANISOTROPIC MODELS

The approach to the singularity in the Robertson-Walker universe is isotropic and the universe collapses to a point. However anisotropic models present a variety of possibilities. Thus Thorne (1967) studied homogeneous models admitting Bianchi type I group of motions and having plane symmetry

$$ds^2 = dt^2 - A^2(dx^2+dy^2) - W^2dz^2$$

where A and W are functions of t alone. The expansions in the x and y directions are identical but different from that in the z direction ($A \neq W$). Thorne found that a dust universe of this type may show two distinct type of behaviour: (i) a collapse in the x-y plane but explosion in the z direction; $A \sim t^{2/3}$, $W \sim t^{-1/3}$ as $t \to 0$, and (ii) pancake collapse; $A \sim (1+at)$, $W \sim t$ as $t \to 0$, so that any domain defined by a finite region of x-y plane retains a finite dimension while the z direction collapses.

The slightly more general metric

$$ds^2 = dt^2 - A^2\,dx^2 - B^2\,dy^2 - C^2\,dz^2$$

with A, B, and C functions of t alone was studied by Raychaudhuri (1958), Heckmann and Schücking (1958), and many others; this exhibit even greater varieties.

A very novel way of approach to the singular state was discovered by Belinskii and Khalatnikov (1969) and studied by Misner (1969b) as well as by the Soviet group (Belinskii *et al.* 1970, 1971, 1972).

The structure constants for the Bianchi type IX are given by $C^1_{23} = C^2_{31} = C^3_{12} = 1$ and the line element can be written in the form

$$ds^2 = dt^2 - l^2(d\psi + \cos\theta\, d\phi)^2 - m^2(\sin\psi\, d\theta - \cos\psi \sin\theta\, d\phi)^2$$

$$-n^2(\cos\psi\, d\theta + \sin\psi \sin\theta\, d\phi)^2 \tag{8.1}$$

where l, m, and n are functions of t alone and $0 \le \psi \le 4\pi$, $0 \le \theta \le \pi$, $0 \le \phi \le 2\pi$. The t lines are here orthogonal to the invariant varieties of the group of motions and would not be the world lines of matter if vorticity were present. Now Shepley (1964) has shown that for this synchronous coordinate system the determinant of the metric tensor vanishes at a finite epoch. (For empty universes, the situation may be different, Misner (1963).) Again in these models as det $|g_{1k}|$ goes to zero, the terms arising from the geometry of the three space (i.e. ${}^3R^a_b$) go to infinity much faster than the energy-stress tensor components if the expansion be anisotropic (Landau and Lifshitz 1971). This is true whether the universe be filled with pressureless dust or ultrarelativistic matter and radiation. Thus, near the singularity, it suffices to consider the empty-space field equations. Indeed the situation may seem paradoxical, that the energy-momentum tensor components, though tending to infinity, have no influence on the dynamics of the universe. For the isotropic universe, however, near the singularity the spatial curvature terms behave as R^{-2} whereas the energy density behaves as R^{-3} (for dust) and R^{-4} (for the ultra-relativistic case). Thus in the isotropic case the curvature terms fall off near the singularity and T terms as well as terms involving the time derivatives $\dot{g}_{ik}$ dominate. This type of singularity has been called the velocity dominated singularity (Eardley, Liang, and Sachs 1972; Liang 1972, 1973) while the curvature dominated singularity as observed by Belinskii and collaborators and Misner is referred to as the mix-master singularity.

With the line element (8.1) the empty-space field equations are (Landau and Lifshitz 1971)

$$\frac{(\dot{l}mn)^{\cdot}}{lmn} = \frac{1}{2l^2m^2n^2}\,[(m^2-n^2)^2-l^4] \tag{8.2a}$$

$$\frac{(l\dot{m}n)^{\cdot}}{lmn} = \frac{1}{2l^2m^2n^2}\,[(n^2-l^2)^2-m^4] \tag{8.2b}$$

$$\frac{(lm\dot{n})^{\cdot}}{lmn} = \frac{1}{2l^2m^2n^2}\,[(l^2-m^2)^2-n^4]\ . \tag{8.2c}$$

$$\frac{\ddot{l}}{l} + \frac{\ddot{m}}{m} + \frac{\ddot{n}}{n} = 0 \tag{8.3}$$

with points signifying differentiation with respect to t. If the r.h.s. of equation (8.2a,b,c) were not present, as is the case with Bianchi type I, one has the Kasner solution

$$l = t^{\mathrm{p}},\ m = t^{\mathrm{q}},\ n = t^{\mathrm{r}} \tag{8.4}$$

$$p+q+r = p^2+q^2+r^2 = 1\ . \tag{8.5}$$

Let us suppose that, even when the terms on the r.h.s. are present, over a certain interval near the singular state we can have the form

$$l \sim t^{\mathrm{p}},\ m \sim t^{\mathrm{q}},\ n \sim t^{\mathrm{r}}, \tag{8.6}$$

then equation (8.3) yields

$$p^2+q^2+r^2 = p+q+r\ . \tag{8.7}$$

Again, at least one of the expressions on the r.h.s. of equations (8.2 a,b,c) must be negative, so that we must have at least one amongst $p.(p+q+r-1)$, $q.(p+q+r-1)$, $r.(p+q+r-1)$ negative. The possibility that p,q, and r are all positive with $(p+q+r-1)$ negative contradicts equation (8.7) and hence we must have at least one of the indices p,q, r negative or in other words length along at least one direction shrinks with time while the spatial volume expands $(p+q+r) > 0$ because of equation (8.7). It is not difficult to see that the equations (8.2a,b,c) do not allow two of the exponents to be negative simultaneously.

To fix our ideas, let us suppose that p is negative with $q < r$. Then in the interval considered m^2 and n^2 can be neglected in comparison with l and the equations (8.2a,b,c) reduce to

$$\alpha'' = -\tfrac{1}{2} e^{4\alpha} \tag{8.8}$$

$$\beta'' = \gamma'' = \tfrac{1}{2} e^{4\alpha} \tag{8.9}$$

where primes denote differentiation with respect to the τ variable

$$d\tau = (lmn)^{-1} dt ,$$

and

$$\alpha \equiv \log l , \quad \beta \equiv \log m , \quad \gamma \equiv \log n \quad .$$

Equation (8.8) indicates a motion of the variable α in an exponential potential well; α' therefore eventually changes sign corresponding to a change from a regime of decreasing l to one of increasing l. Let us suppose, with Belinskii et al. that at a certain epoch, the r.h.s. of equations (8.2a,b,c) are all small enough so that $p+q+r \approx 1$ and we have the Kasner solution with

$$lmn = \Lambda t, \quad \tau = \Lambda^{-1} \ln t \; + \text{const.}$$

where Λ is a constant. Now equations (8.8) and (8.9) may be integrated to give

$$l^2 = 2|p|\Lambda[\cosh(2p\Lambda\tau)]^{-1} ,$$

$$m^2 = m_0^2 \exp[2\Lambda q\tau] \times [1+\exp(4p\Lambda\tau] ,$$

$$n^2 = n_0^2 \exp[2\Lambda r\tau] \times [1+\exp(4p\Lambda\tau)] ,$$

where the integration constants have been so chosen that as $\tau\to\infty$ and l, m and n go to the assumed Kasner solution with a negativ

p. The asymptotic values of l,m,n as τ tends to $+\infty$ and $-\infty$ are

$$l \sim e^{p\Lambda\tau}, \quad m \sim e^{q\Lambda\tau}, \quad n \sim e^{r\Lambda\tau}, \quad \Lambda\tau \sim \ln t \quad \text{as } \tau \to +\infty$$

$$l \sim e^{-p\Lambda\tau}, \quad m \sim e^{\Lambda(q+2p)\tau}, \quad n \sim e^{\Lambda(r+2p)\tau}, \quad \Lambda(1+2p)\tau \sim t$$

$$\text{as } \tau \to -\infty \, .$$

In the latter limit, transforming $\tau \to t$, we get

$$l \sim t^{p'}, \quad m \sim t^{q'}, \quad n \sim t^{r'}$$

with

$$p' = -\frac{p}{1+2p} > 0 \, ,$$

$$q' = \frac{2p+q}{1+2p} < 0 \, ,$$

$$r' = \frac{r+2p}{1+2p} > 0 \, .$$

The situation has now changed from that prevailing at $\tau \to +\infty$ in the sense that the exponent of l has changed to positive values and that of m has become negative. Thus the l,m axes have interchanged their contracting and expanding characters. We conclude that the evolution of the metric as we proceed towards the singularity is made up of 'successive periods (called eras) during which distances along two of the axes oscillate while distances along the third axis decrease. On going from one era to the next, the direction along which distances decrease monotonically bounces from one axis to another. The order of this bouncing acquires asymptotically the character of a random process' (Landau and Lifshitz 1971.) It may be further shown that one has a specially long era if in any case p,q,r corresponds to the triplet (100). Now these values are especially interesting because in these cases there are no particle horizons in the direction for which the index is unity as $\int_0 t^{-1}\,dt$ diverges. During the long eras, therefore, there occurs a mixing up in this direction and, as in the course of evolution this particular direction also changes, one may expect a mixing up and effective abolition of all particle horizons. This was the idea of the mix master model of Misner but as we have already noted it did not succeed in providing an explanation of the apparent breakdown of causality or of the evolution of an isotropic universe from one initially showing serious deviations from isotropy.

We have reproduced above some of the mathematical discussions given by Belinskii *et al.*; identical conclusions have been reached by Misner (1969b) with a different mathematical procedure. Misner has also emphasized that while the t time to the singularity is finite, the τ time is infinite. Near the singularity, τ is physically significant as it is a measure of the time taken for light to circumnavigate the universe. Thus, in a sense, the singularity may be said to be in the infinite past for infinitely many times light has gone round the universe since the singularity.

9

THE GRAVITATIONAL CONSTANT AS A FIELD VARIABLE

9.1. THE LARGE NON-DIMENSIONAL NUMBERS AND THE DIRAC HYPOTHESIS

Considering the cosmological and the microphysical data, one can build up large non-dimensional numbers which show a surprising closeness between them. Thus

(a) $$\frac{cH_0^{-1}}{(e^2/m_e c^2)} \approx 6 \times 10^{40}$$

(b) $$\frac{e^2}{Gm_p m_e} \simeq 0 \cdot 2 \times 10^{40}$$

(c) $$\frac{\rho_0 (cH_0^{-1})^3}{m_p} \approx 10^{80} = (10^{40})^2$$

where e is the electronic charge, m_p and m_e the masses of the protons and electrons respectively, and ρ_0 is the density of matter in the universe at present. The quantities (a), (b) are respectively the ratio of the 'Hubble length' and the classical electronic-radius and the ratio of electrostatic and gravitational interactions between a proton and an electron, while (c) may be looked upon as the number of nucleons in the universe. Of course there is some uncertainty in the value of H_0 and a good deal more in ρ_0 but such uncertainties cannot invalidate the essential closeness of the numbers involved. The microwave background radiation has brought to our notice another large number; the ratio of the number of photons to baryons or equivalently the entropy per baryon. This non-dimensional number is not far from $(10^{40})^{\frac{1}{4}}$.

Is this apparently strange agreement between large non-dimensional numbers purely accidental? In the light of our present-day physics, it is indeed so, for the Hubble constant

H_0 and the cosmic density would vary with time while the others (e, G, m_p, m_e, c) remain constant and so the agreement will not hold at other epochs. The agreement is, however, so impressive that Dirac (1938, 1974) sought to change some basic ideas of physics so that this agreement would be a permanent feature of the universe. He introduced thefollowing postulate as a fundamental principle: 'Any two of the very large non-dimensional numbers occurring in nature are connected by a simple relation, in which the coefficients are of the order unity'.

On the basis of the Dirac postulate the ratio of the gravitational force and the electrostatic force must vary with epoch to maintain the agreement between (a) and (b). This would mean that G,e, and the masses of the fundamental particles cannot all be constants. Instead of building up a formalism in which one or more of these would be clearly variable requiring a change in some of the current physical theories, Dirac tried to make a compromise. He considered two physics - the micro or atomic physics where e, m, and c are by definition constants (i.e. they define the standards of length, mass, and time) and G varies with time being inversely proportional to the age of the universe. (The age of the universe being H^{-1} up to a factor of the order of unity.) In the other physics - the macro or general relativity physics - m, c, and G are by definition constants (these now define the standards) and the charge e is variable. However as Bondi (1962a) points out the constancy or conservation of charge is an integral part of the Maxwell equations and is demanded by both the special and the general theory of relativity.

9.2. A VARYING G

Before going to the formal framework of theories involving a variation of G (Jordan 1955) let us report some other arguments that have been advanced in favour of a varying G and also of the observational evidences on this point. Sciama considers a particle in accelerated motion in the neighbourhood of a massive body. He argues that on the Mach principle the particle may as well consider the universe to be in motion with an acceleration in the opposite direction. This accelerated motion is supposed to generate a field in a manner similar to the field due to a charged particle in Maxwell's theory. Thus Sciama (1953) writes for this Machian field

$$\phi = -\int \frac{\rho d\tau}{r}, \quad \underline{A} = -\int \frac{\rho \underline{v}}{cr} d\tau .$$

The integrals would obviously diverge if taken over infinite space. However one must remember the particle horizon in big-bang models due to which at any instant influences have reached only up to a finite distance. Therefore taking the integrals over a spherical region of radius cT (T being the age of the universe) one obtains

$$\phi \sim -2\pi\rho c^2 T^2 , \quad \underline{A} \sim \phi\frac{\underline{v}}{c} .$$

For the homogeneous universe ϕ is thus spatially constant and one obtains for the 'field intensity'

$$\underline{E} \sim -\frac{\phi}{c^2}\frac{\partial \underline{v}}{\partial t}$$

as ϕ is a very slowly varying function as compared with $\underline{v}$. Again for the field due to the massive body in the neighbourhood we similarly write

$$\underline{E}' = -\frac{M}{r^3}\underline{r} + \frac{M}{c^2 r}\frac{\partial \underline{v}}{\partial t} .$$

In the rest frame of the particle the total field must vanish:

$$-\frac{M}{r^3}\underline{r} + \frac{M}{c^2 r}\frac{\partial \underline{v}}{\partial t} - \frac{\phi}{c^2}\frac{\partial \underline{v}}{\partial t} = 0 . \tag{9.1}$$

If now we write the Newtonian equation of motion

$$\frac{\partial \underline{v}}{\partial t} = -\frac{GM}{r^3}\underline{r} \tag{9.2}$$

then comparing equations (9.1) and (9.2) we get

$$\frac{1}{G} = \left(-\phi + \frac{M}{r}\right)\frac{1}{c^2} \sim -\frac{\phi}{c^2} \sim \rho T^2 ,$$

i.e.

$$G\rho T^2 \sim 1 \ . \tag{9.3}$$

Dicke (1962b) has pointed out that the relation (9.3) can be obtained from simple dimensional considerations. According to Mach's principle the universe is to play a role in the acceleration of a particle in the vicinity of a gravitating body (e.g. the acceleration of a planet due to the sun). So the matter density ρ and the age of the universe may come into the expression of this acceleration. Now using Newton's law of gravitation, one gets for the acceleration f

$$f \approx \frac{M}{r} \cdot \frac{1}{\rho T^2}$$

and the relation (9.3) follows.

Looking to the observational side, it is obvious that a variation of G would have consequences in the field of geophysics; indeed phenomena like continental drift have been cited as evidences in favour of a decreasing G but there are so many complicating factors that geophysics can hardly be expected to give any convincing information about the constancy or otherwise of G.

Teller (1948) made an estimate of the thermal history of the earth which, with the then prevailing value of the Hubble constant, seemed to indicate that a variation of G at the rate demanded by Dirac's postulate did not occur. From the virial theorem one obtains for the temperature T of a star of mass M and radius R

$$T \propto GMR^{-1} \ .$$

Again the luminosity of the star is proportional to the product of the radiation-energy gradient ($\sim\sigma T^4 R^{-1}$), the mean free path of photons λ and surface area R^2:

$$L \propto RT^4\lambda \ .$$

Astrophysical calculation gives $\lambda \propto T^3 R^6 M^{-2}$, so that combining all these $L \propto G^7 M^5$.

Again considering the orbit of the earth to be a circle of radius r, the linear velocity v satisfies the relation $v^2 = GMr^{-1}$. The force being central, the angular momentum about the centre mvr will be conserved, hence $r \propto (GM)^{-1}$. The decrease of G thus leads to an expansion of the orbit. Applying Stefan's law the temperature of the earth would vary as the fourth root of the energy received, i.e. $(Lr^{-2})^{\frac{1}{4}}$. Thus finally one gets for the temperature of the earth's surface T_E:

$$T_E \propto G^{2\cdot 25 \;\; 1\cdot 75} \quad .$$

With the Dirac hypothesis $G \propto T^{-1}$, and the then prevailing estimate of the age of the universe ~ 2×10^9 years, Teller concluded that at 2 to 3 hundred million years ago, G was almost 10 per cent greater and consequently T_E was near the boiling point of water. The known existence of dinosaurs at these times contradicts such a high temperature. However in the above calculation no account has been taken of the albedo of cloud in controlling the earth's surface temperature and besides the presently accepted age of the universe is ~10^{10} years, so that the temperature T_E would have been only a few degrees above the present surface temperature of the earth. Again as we shall see, the variation of G in the Brans-Dicke theory may be much slower, so that even 2×10^9 years ago, the temperature would have been at most some 10° to 20° higher than at present. This, instead of standing in the way of biological evolution, may have rather provided congenial conditions for the origin of life from inanimate matter at an earlier epoch (Dicke 1962b).

Some years ago Pochoda and Schwarzschild (1964) examined the effect of the higher luminosity (i.e. a greater hydrogen burning) consequent on a higher value of G in the past on the evolution of the sun. They examined a variation of the form

$$G_t = G_0 \left(\frac{T}{t}\right)^n$$

where G_0 is the present value of G (i.e. at $t = T$). They found that if $n = 1$ (Dirac's hypothesis) then the sun should have been a red giant by now unless $T \geq 1{\cdot}5\ 10^{10}$ years.

Recently there has been an attempt to determine directly the variation of G from an analysis of radar-echo time delays between the earth and mercury. In effect, Shapiro, Smith, Ingalls, and Pettengill (1971) were studying the effect of a possible variation of G on planetary motion. They concluded $(\dot{G}/G)_0 < 4{\times}10^{-10}\ \mathrm{yr}^{-1}$, the subscript 0 indicating the value at present. An expectation was expressed that continuation of the observations for five more years would increase the accuracy by an order of magnitude, i.e. a value of $(\dot{G}/G) \sim 3{\times}10^{-11}\ \mathrm{yr}^{-1}$ would be detectable - so far, however, these results have not appeared. Morganstern (1972) points out that a negative result even of that order would not be a conclusive evidence against the Brans-Dicke theory as such but would merely eliminate some special models of Brans-Dicke universes. From the observed spin down of pulsars and angular momentum conservation, Mansfield (1976) gives $\dot{G}/G \leq 5{\cdot}8 \pm 1{\times}10^{-11}\ \mathrm{yr}^{-1}$.

Dearborn and Schramm (1974, cf. Morrison 1973) from a study of the cluster of galaxies have concluded

$$\dot{G}/G \leq 4{\times}10^{-11}\ \mathrm{yr}^{-1} \quad \text{if } q_0 \ll 1$$

$$\dot{G}/G \leq 6{\times}10^{-11}\ \mathrm{yr}^{-1} \quad \text{if } q = 1/2$$

with $H_0 = 55\ \mathrm{km\ s^{-1}\ Mpc^{-1}}$ in either case. However even these may not be inconsistent with the Brans-Dicke theory, but apparently the theory of Hoyle and Narlikar (1972) is contradicted. (See however, Marchant and Mansfield, 1977.)

Chen and Stothers (1976) conclude from a study of solar evolution that $(\dot{G}/G) < 1{\times}10^{-10}\ \mathrm{yr}^{-1}$. The only positive result so far seems to be that due to Van Flandern (1974, 1975) who attributes an apparently unaccounted residual secular acceleration of the moon to a variation of G with the present value of $(\dot{G}/G) = (-8{\pm}5){\times}10^{-11}\ \mathrm{yr}^{-1}$ (Heintzmann and Hillerbrandt 1975).

9.3. THE SCALAR TENSOR THEORY OF BRANS AND DICKE

The problem is to introduce a varying G in the formalism of field theory. One brings in besides the metric tensor a new scalar field-variable. This scalar must not however be one derived from the invariants of Riemannian geometry for they fall off more rapidly than $1/r$ and consequently are inadequate to bring in any Machian influence of distant matter. Further, Brans and Dicke like to maintain the energy momentum conservation as well as the geodesic nature of the paths. They are thus led to the variational principle

$$\delta \int [\phi R + \frac{16\pi L}{c^4} - \omega\phi_i\phi^i\phi^{-1}]\, (-g)^{\frac{1}{2}} \mathrm{d}x^0 \mathrm{d}x^1 \mathrm{d}x^2 \mathrm{d}x^3 = 0 \, , \tag{9.4}$$

where ϕ is the new scalar field and $\phi_{,i} \equiv \phi_i$ and is the matter Lagrangian so that

$$T^{ij} = + \frac{2}{(-g)^{\frac{1}{2}}} \frac{\partial}{\partial g_{ij}} \sqrt{(-g)}\, L.$$

The variation is to be performed regarding g_{ik} and ϕ as independent variables. It is interesting to compare the Lagrangian of Brans and Dicke with that of general relativity. The Einstein equations follow from the variational principle

$$\delta \int [R + 16\pi G c^{-4} L] (-g)^{\frac{1}{2}} \mathrm{d}x^0 \mathrm{d}x^1 \mathrm{d}x^2 \mathrm{d}x^3 = 0 \, .$$

Thus in the Lagrangian of Brans and Dicke the coupling constant G of general relativity has been replaced by ϕ^{-1} and instead of attaching it to the matter Lagrangian it is affixed to the geometric scalar R. This effectually means that the equations of motion and the conservation relations are not disturbed even though the coupling ϕ is now a field variable. The third term is the typcial Lagrangian for a scalar field, the rest mass being taken as zero to have a long-range field. The coupling constant ω is left undetermined which is an obvious weak point of the theory although it gives some scope for manipulation when questions of fit with observational data are taken up.

The Lagrangian introduced by Jordan in his theory is

somewhat more general - the variational principle being

$$\delta \int \chi^{\eta}(R + \zeta\chi^{-2}\chi_i\chi^i + \chi L)(-g)^{\frac{1}{2}}\, dx^0 dx^1 dx^2 dx^3 = 0 \ .$$

It is apparent that if the constant $\eta = -1$, then with the identification $\chi^{-1} = \phi$, Jordan's theory goes over to that of Brans and Dicke with $\zeta = -\omega$. However, for more general values of η, neither the conservation principle nor the geodesic nature of paths hold good (Brill 1962). There are however other scalar tensor theories as well (e.g. Schwinger 1968; Milton and Ng 1974) and Harrison (1972a) has given an illuminating discussion on the relation between different scalar-tensor theories and general relativity.

Returning to (9.4), the field equations of Brans and Dicke (1961) are

$$R_{ij} - \tfrac{1}{2} g_{ij}R = -\frac{8\pi\phi^{-1}}{c^4} T_{ij} - \frac{\omega}{\phi^2}(\phi_i\phi_j - g_{ij}\phi_k\phi^k)$$

$$- \frac{1}{\phi}(\phi_{i;j} - g_{ij}\Box\phi) \tag{9.5}$$

$$\Box\phi \equiv g^{ik}\,\phi_{i;k} = \frac{8\pi}{(3+2\omega)c^4}\, T \tag{9.6}$$

where the scalar T is the trace of the tensor T^i_k and the conservation relation for the energy-momentum tensor follows identically

$$T^{ik}_{;k} = 0 \ .$$

Equation (9.6) would give for a static distribution of particles

$$G^{-1} = \phi \sim \Sigma \frac{m}{rc^2}$$

which is a Machian relation showing the influence of distant matter. Brans and Dicke gave an approximate solution of the equations for weak fields with spherical symmetry

$$\phi = \phi_0 + 2M(3+2\omega)^{-1}c^{-2}r^{-1},$$

$$g_{00} = 1-2M\phi_0^{-1}r^{-1}c^{-2}[1+(3+2\omega)^{-1}] ,$$

$$g_{ik} = -1-2M\phi_0^{-1}r^{-1}c^{-2}[1-(3+2\omega)^{-1}] \quad \text{for } i=k\neq 0 \ .(i,k = 1,2,3) .$$

$$g_{\alpha\beta} = 0 \quad \text{if} \quad \alpha\neq\beta . \quad (\alpha,\beta = 1,2,3,0) .$$

The expression for g_{00} assumes the general relativity form, if one defines the asymptotic value of the gravitational constant as

$$G_0 = \phi_0^{-1}(4+2\omega)(3+2\omega)^{-1} ,$$

when the gravitational red shift will have the Einstein value. One can calculate the deflection of light with this weak field solution

$$\delta\theta = 4G_0Mr^{-1}c^{-2}(3+2\omega)(4+2\omega)^{-1}$$

which differs from the relativity expression due to the term involving the coupling constant ω. Indeed as $\omega\to\infty$, the Brans-Dicke value goes over to the general relativity value. As, however, the deflection of light cannot be determined to a very high degree of accuracy, the expression for $\delta\theta$ has not been able to decide between the Brans-Dicke theory and the general theory of relativity so far.

It is however possible that radio-astronomical observations will be able to determine $\delta\theta$ to a sufficiently high degree of accuracy. Thus the deflection of radio waves in the sun's gravitational field has been measured by observing the radio source 3C279 before and after its occultation by the sun. The observational results are generally in somewhat better agreement with the Einstein value but are by no means decisive (Seilestad, Sramek, and Weiler 1970; Muhleman, Ekers, and Fomalont 1970; Sramek 1971; Riley 1973; Shapiro, Ash, Ingalls, Smith, Campbell, Dyce, Jurgons, and Pettengill (1971). (See also Jones, B.F. (1976) and Weiler, Ekers, Raimond, and Wellington 1974).

The situation is somewhat different for the perihelion motion of mercury where the observational data are much more accurate. In a discussion of perihelion motion, it is necessary to consider the solution to a higher order of approximation than the weak field approximation given above. Brans and Dicke (see also Brans (1962)) gave the exact solution for a spherically symmetric vacuum field as follows

$$ds^2 = e^{2\alpha}dt^2 - e^{2\beta}(dr^2+r^2d\Omega^2)$$

$$e^{2\alpha} = [(1-Br^{-1})(1+Br^{-1})^{-1}]^{2/\lambda}$$

$$e^{2\beta} = (1+Br^{-1})^4[(1-Br^{-1})(1+Br^{-1})^{-1}]^{2(\lambda-C-1)/\lambda}$$

$$\phi = \phi_0[(1-Br^{-1})(1+Br^{-1})^{-1}]^{C/\lambda}$$

$$\lambda = \left[(C+1)^2 - C\left(1 - \frac{\omega C}{2}\right)\right]^{\frac{1}{2}}$$

$$C = -(2+\omega)^{-1}$$

$$B = M(2c^2\phi_0)^{-1}[(2\omega+4)(2\omega+3)^{-1}]^{\frac{1}{2}}$$

$$G_0 = \phi_0^{-1}(4+2\omega)(3+2\omega)^{-1} \quad .$$

The solution goes over to the Schwarzschild metric as $\omega\to\infty$ (when $C \to 0$, $\lambda \to 1$, $\phi \to \phi_0 \to G^{-1}$, $B \to GM/2c^2$). With this metric the perihelion motion comes out as

$$\frac{4+3\omega}{6+3\omega} \times \text{Einstein value.}$$

It is generally believed that the Einstein value is in excellent agreement with the observed perihelion motion of mercury and in that case one has to set a fairly high value to ω in order not to run into contradiction with observation (say $\omega \approx 20$). (Of course the simpler thing would be to forget ω altogether and accept general relativity in good grace.) However Dicke and Goldenberg (1967) claimed to have discovered that the sun is an oblate spheroid, the fractional difference of equatorial and polar radii $(r_{eq} - r_{pole})/r$ having a value $5{\cdot}0 \pm 0{\cdot}7{\times}10^{-5}$, (quadrupole moment $\sim 5{\times}10^{-5}$). This would account for a perihelion motion of mercury of 3·4" per century leaving an unexplained motion of 39·6" per century. This is about 8 per cent less than the Einstein value. One can then have a value of ω between 4 and 7 (say). However the question of solar oblateness remains controversial and in general the scientific community has not accepted the interpretation of Dicke and Goldenberg (Durney and Werner 1971; Ingersoll and Spiegel 1971; Chapman and Ingersoll 1972, 1973; Chapman 1975; Hill and Stebbins 1975).

Dicke (1962a) has also given an alternative version of their theory by a simple redefinition of units. This makes the scalar field appear as a non-gravitational field, the Einstein equations in their original form being satisfied but the usual energy-momentum tensor is augmented by that due to the scalar field. In this new version G, h, and c do not vary but the masses of elementary particles become dependent on their space-time position. Thus the 'gravitational' red shift is now only partly due directly to the metric, the other part arising from a change in the energy levels due to change of the electronic mass. While the theory in this form may appear somewhat less appealing than the previous version, the equations in the new form are a little simpler. However, the paths for free fall, except those of massless particles, are no longer geodesics and the conservation principle holds only for the combined energy-tensor due to matter and the scalar field. Thus if the units of time, length, and reciprocal mass are scaled by a factor $\lambda^{-\frac{1}{2}}$, so that $g_{ik} \rightarrow \lambda g_{ik}$ etc., one obtains the field equations (with $\lambda = \phi/\bar{\phi}$, $\bar{\phi}$ being a constant)

$$R_{ij} - \tfrac{1}{2}g_{ij}R = -\frac{8\pi G}{c^4}(T_{ij}+\Lambda_{ij}) \,, \tag{9.7}$$

$$\Lambda_{ij} = \frac{2\omega+3}{16\pi G\lambda^2}(\lambda_i\lambda_j - \tfrac{1}{2}g_{ij}\lambda_k\lambda^k) \,, \tag{9.8}$$

$$\Box(\log\lambda) = \frac{8\pi GT}{c^4(2\omega+3)} \,, \tag{9.9}$$

where T_{ij} is the usual energy-momentum tensor with masses varying as $\lambda^{-\frac{1}{2}}$, i.e. $m = m_0\lambda^{-\frac{1}{2}}$.

An important difference in the spherically symmetric case between the Brans-Dicke theory and the general theory of relativity should be noted. In general relativity the Schwarzschild solution is the unique solution for spherically-symmetric vacuum, so that there cannot be spherically symmetric gravitational waves, and further the predictions about the three crucial tests are also unique. For the Brans-Dicke theory this is not so - an almost trivial way of seeing that is to note that the Schwarzschild metric with ϕ = const. is also a solution for the vacuum case. In the non-static case there also exist wave solutions. The particular solutions presented by Brans and Dicke imply, beside the field equations, some boundary conditions which may involve Machian ideas. It is of some interest to note that there are a number of vacuum solutions with the Robertson-Walker metric (O'Hanlan and Tupper 1972).

Excluding the case of vacuum and distribution of massless particles or electromagnetic fields (i.e. if $T \neq 0$), ϕ is not a constant and thus a general relativity solution is not a solution of the Brans-Dicke equations. In particular for stationary homogeneous cosmological models (i.e. where the space-time admits a transitive group of motions) one may argue that the scalar ϕ must be a constant, hence there will be no solution of Brans-Dicke equations for $T \neq 0$. Thus it has been claimed that homogeneous universes in rigid rotation are not allowed in Brans-Dicke theory (Banerji 1974).

9.4. COSMOLOGICAL SOLUTIONS OF THE BRANS-DICKE THEORY

We shall seek solutions of the Brans-Dicke equations subject to the following conditions: (i) The cosmological principle is valid so that the line element is of the Robertson-Walker form, (ii) The scalar ϕ (or λ) is a function of t alone.

McIntosh (1973) has shown that in units in which G vary there are solutions of the Brans-Dicke equations with the Robertson-Walker line element and with ϕ a function of both r and t. In these solutions the matter density ρ is also a a function of the spatial co-ordinates.

Using equations (9.7) - (9.9), we shall investigate the function R in the Robertson-Walker line element. The only non-vanishing components of Λ_{ij} are

$$-p_\lambda \equiv \Lambda_3^3 = \Lambda_2^2 = \Lambda_1^1 = -\frac{2\omega+3}{32\pi G}\left(\frac{\dot\lambda}{\lambda}\right)^2 ,$$

$$\rho_\lambda \equiv \Lambda_0^0 = \frac{2\omega+3}{32\pi G}\left(\frac{\dot\lambda}{\lambda}\right)^2 .$$

Thus effectively besides the matter field there is another 'perfect fluid' field with $p_\lambda = \rho_\lambda$. The field equations are thus

$$\frac{8\pi G}{3}(\rho+\rho_\lambda) = \frac{k}{R^2} + \frac{\dot R^2}{R^2} , \tag{9.10}$$

$$-8\pi G(p+p_\lambda) = \frac{k}{R^2} + \frac{\dot R^2}{R^2} + \frac{2\ddot R}{R} , \tag{9.11}$$

so that

$$\frac{\ddot R}{R} = -\frac{4\pi G}{3}(\rho+3p+4\rho_\lambda) . \tag{9.12}$$

With $\omega > -3/2$, ρ_λ is non-negative and hence with ρ, p non-negative one has necessarily a singular state of $R = 0$ at a finite time in the past. The positive definite energy of the scalar field simply hastens the collapse.

One may write the conservation relations in two different forms. From equations (9.11) and (9.10) we have as in relativistic fluid dynamics:

$$\frac{d}{dt}[(\rho+\rho_\lambda)R^3] + [3(p+\rho_\lambda)R^2\frac{dR}{dt}] = 0 . \tag{9.13}$$

However, taking directly the divergence of equation (9.7), we get

$$\frac{d}{dt}(\rho R^3) + 3pR^2\frac{dR}{dt} + \tfrac{1}{2}R^3\frac{\dot\lambda}{\lambda}(\rho-3p) = 0 , \tag{9.14}$$

where we have used equation (9.9). The divergence relation also confirms that in the present case with p, ρ, and λ functions of t alone, the fluid-velocity vector is geodetic. Equation (9.9) written out in full reads in this case

$$\left(\frac{\dot{\lambda}}{\lambda} R^3\right)^{\cdot} = \frac{8\pi R^3 G}{(2\omega+3)} (\rho - 3p) . \tag{9.15}$$

Equations (9.10), (9.13), (9.14), and (9.15) are obviously not independent; we shall take (9.10), (9.7), and (9.8) as the equations of our problem. The equation system has been integrated for all curvatures (k=0,±1) in case of a pure radiation universe (Morganstern 1971a,b) and in case $k = 0$ for a dust universe (Dicke 1968). (See also Bishop 1976.)

Let us consider the dust universe with $k = 0$. Equation (9.14) integrates to

$$\rho R^3 \lambda^{\frac{1}{2}} = \text{const.} \tag{9.16}$$

This equation could have been written down directly from the fact that in Dicke's revised version, masses are not constants but vary as $\lambda^{-\frac{1}{2}}$. Equations (9.13) and (9.14) together give for $p = 0$

$$(\rho_\lambda R^3)^{\cdot} + 3\rho_\lambda R^2 \dot{R} - \tfrac{1}{2}\rho R^3 \frac{\dot{\lambda}}{\lambda} = 0 \tag{9.17}$$

or

$$(\rho_\lambda R^6)^{\cdot} = \tfrac{1}{2} \rho R^6 \frac{\dot{\lambda}}{\lambda} .$$

Using equation (9.10) with $k = 0$ and replacing $\dot{\lambda}/\lambda$ by ρ_λ we get from (9.17)

$$\frac{\mathrm{d}}{\mathrm{d}R} [\rho_\lambda R^6]^{\frac{1}{2}} = \pm \left(\frac{3}{2\omega+3}\right)^{\frac{1}{2}} \cdot \frac{\rho R^2}{(\rho+\rho_\lambda)^{\frac{1}{2}}} . \tag{9.18}$$

Again from equation (9.13)

$$\frac{\mathrm{d}}{\mathrm{d}R} [(\rho+\rho_\lambda)R^6] = 3R^5 \rho , \tag{9.19}$$

so that combining equations (9.18) and (9.19), we get

$$\frac{d}{dR}\,[\rho_\lambda R^6]^{\frac{1}{2}} = \pm\frac{1}{[3(2\omega+3)]^{\frac{1}{2}}}\,\frac{d}{dR}\,[(\rho+\rho_\lambda)R^6]^{\frac{1}{2}} , \tag{9.20}$$

which, on integration, gives

$$(\rho_\lambda R^6)^{\frac{1}{2}} = \pm\left[\frac{(\rho+\rho_\lambda)R^6}{6\omega+9}\right]^{\frac{1}{2}} + C . \tag{9.21}$$

A very simple type of solution is obtained if the integration constant C vanishes

$$\rho = (6\omega+8)\rho_\lambda . \tag{9.22}$$

Finally we have

$$\begin{aligned} \rho &= \rho_0 \left(\frac{R_0}{R}\right)^{(6\omega+10)/(2\omega+3)} , \\ R &= R_0 \left(\frac{t}{t_0}\right)^{(2\omega+3)/(3\omega+5)} , \\ \lambda &= \lambda_0 \left(\frac{t}{t_0}\right)^{2/(3\omega+5)} , \end{aligned} \tag{9.23}$$

where the zero of time has been chosen at the singular state.

Now in this version the mass, length, and time are changed by the factor $\lambda^{\frac{1}{2}}$ whereas G remains constant. To have the change of G in more conventional units, we have, using capital letters for the quantities in the conventional units:

$$m = M\lambda^{-\frac{1}{2}}$$

$$t = T\lambda^{\frac{1}{2}}$$

$$l = L\lambda^{\frac{1}{2}}$$

$$\lambda = \lambda_0 \left(\frac{T}{T_0}\right)^{2/(3\omega+4)} \qquad \text{or } \phi = \phi_0 \left(\frac{T}{T_0}\right)^{2/(3\omega+4)}$$

$$R = R_0 \left(\frac{T}{T_0}\right)^{2(1+\omega)/(3\omega+4)} \tag{9.24}$$

$$\rho = \rho_0 \left(\frac{R_0}{R}\right)^3 = \rho_0 \left(\frac{T}{T_0}\right)^{-6(1+\omega)/(3\omega+4)}$$

where R and ρ are now in mass-constant units and so R is now $\lambda^{-\frac{1}{2}}$ times the R of equation (9.23). Further in view of equation (9.6) we get

$$\phi_0 = 4\pi . \frac{3\omega+4}{(2\omega+3)} \rho_0 R_0^3 T_0^2 . \tag{9.25}$$

Equation (9.24) shows that in this case ($p = 0$, $k = 0$, $C = 0$) ϕ increases monotonically being zero at the singularity and arbitrarily large as R and T go to infinity. Thus G decreases monotonically. However with $\omega \approx 5$, $G \propto T^{-0 \cdot 1}$ and is thus not consistent with the Dirac idea $G \sim T^{-1}$. Therefore in this form of the model, the agreement between the large non-dimensional numbers would not persist at different epochs. The softer variation of G has however the desirable consequence that no direct contradiction with observations occurs.

When C is not zero, the situation is much more complicated, although one can still integrate the equations in closed form (Greenstein 1968a,b; Weinberg 1972b). However one can gain considerable insight from the following simple considerations: (i) As $t \to \infty$, in view of equation (9.10) R will be arbitrarily large and obviously (9.23) or equivalently (9.24) will be an asymptotic solution for all C. (ii) At the other limit, i.e. as $t \to 0$, $R \to 0$. Let us consider the case $\rho_\lambda \gg \rho$. Then we have from equation (9.21) $\rho_\lambda \propto R^{-6}$, and hence from equation (9.10) $R \sim t^{1/3}$. Also, remembering the expression for ρ_λ,

$$\left(\frac{2\omega+3}{12}\right)^{\frac{1}{2}} \frac{\dot{\lambda}}{\lambda} = \pm \frac{\dot{R}}{R}$$

or

$$\lambda \sim R^{\pm(12/(2\omega+3))^{\frac{1}{2}}}.$$

Obviously the situation $\rho_\lambda \gg \rho$ can obtain only if C is positive (see 9.21). Also we have from equation (9.16)

$$\rho \sim R^{-3\pm(3/(2\omega+3))^{\frac{1}{2}}}$$

which is consistent with $\rho \ll \rho_\lambda$. In the mass-constant units, ths solution is

$$\begin{aligned} t &= T^{(1\pm\alpha/3)^{-1}} \\ \lambda &\sim T^{\pm 2\alpha/(3\pm\alpha)} \\ R &\sim T^{(1\pm\alpha)/(3\pm\alpha)} \\ \alpha &\equiv + \left(\frac{3}{2\omega+3}\right)^{\frac{1}{2}} . \end{aligned} \tag{9.26}$$

Thus with $C \neq 0$, we can have both a decreasing and increasing G at early stages, although for large t and R all the solutions behave in a similar manner.

For $k = 0$, the integration can be carried out for an equation of state $p = \varepsilon\rho$, where ε is a constant. The case $\varepsilon = 1/3$ corresponding to radiation and ultrarelativistic gas is specially interesting as this is the condition obtaining in the early stages of the hot big-bang models. The scalar field is now source free, so that the solutions of general relativity satisfy the Brans-Dicke equations as well. Some other solutions for this case have been given by Morganstern (1971b). Equations (9.14) and (9.15) give

$$\begin{aligned} \frac{\dot{\lambda}}{\lambda} R^3 &= \text{const.} \\ \rho R^4 &= \text{const.} \end{aligned} \tag{9.27}$$

Utilizing these relations Morganstern is able to integrate (9.10) and finds for all k as $R \to 0$,

$$R \propto t^{1/3}$$

$$\lambda \propto t^{2/[(6\omega+9)^{\frac{1}{2}}]} \, . \tag{9.28}$$

The scalar-field energy dominates near $R = 0$: $\rho_\lambda \sim R^{-6}$ and $\rho \sim R^{-4}$. However with p slightly different from $\rho/3$ (e.g. $p = 0{\cdot}3\rho$) one can have solutions in which near the singularity ρ_λ is significantly less than ρ ($\rho_\lambda = 0{\cdot}0075\rho/(2\omega+3)$) (Dicke 1968). In the limit these solutions pass over to the general relativity solutions.

9.5. COMPARISON WITH OBSERVATIONS

The crucial point of the Brans-Dicke theory is the variation of G and we have already given a discussion of some of the observational and calculational results in this connection. We have also seen that so far observations have not been able to decide between the Brans-Dicke results and the Einstein values on the three crucial tests. In cosmology, one may seek to make a study of the m-z, number count-magnitude, and the angular size measurements against the background of this theory. However the variety of models and consequently the arbitrary parameters are even greater than in relativistic cosmology. In most cases one rather arbitrarily takes $p = k = C = 0$ solution for comparisons just as in general relativity the Einstein-de Sitter model has been most intensively studied because of its simplicity. While the Einstein-de Sitter model requires the deceleration parameter q to be $\frac{1}{2}$, for the solution (9.24) of the last section $q = (\omega+2)(2+2\omega)^{-1}$ which for $\omega \approx 5$, gives $q \approx 0{\cdot}58$. Again the age, instead of being $2H_0^{-1}/3$, is in this case $H_0^{-1}(2\omega+2)(3\omega+4)^{-1}$, i.e. $\approx 0{\cdot}63\ H_0^{-1}$, thus the age is slightly decreased apparently due to the gravitational influence of the scalar field energy. However a greater G in the past would mean a greater luminosity of stars and thus the distant galaxies are actually at distances greater than the values assigned on the basis of a constant absolute luminosity. This thus brings in another new evolutionary correction in the determination of H_0 and q_0 (Tinsley 1972b).

Luke and Szamosi (1972) have made an integration of the

Brans-Dicke field equations by a self consistent numerical method. The field equations for the Robertson-Walker metric are reduced to the form (with $p = 0$ and with 9.5 and 9.6)

$$\frac{\dot{R}^2}{R^2} = -\frac{k}{R^2} + \frac{8\pi\rho}{3\phi} + \frac{\omega\dot{\phi}^2}{6\phi^2} - \frac{\dot{\phi}\dot{R}}{\phi R}$$

$$(R^3\dot{\phi})^{\cdot} = \frac{8\pi\rho R^3}{(3+2\omega)} = \text{const.}$$

Luke and Szamosi obtain a lower bound on $|(\dot{G}/G)|$ as 8×10^{-13} yr^{-1}. Combining this with the limit set by Shapiro, Smith, Ingalls and Pettengill (1971), they give

$$3{\cdot}0 \times 10^{-11}\text{yr}^{-1} > \left|\left(\frac{\dot{G}}{G}\right)\right| > 8{\cdot}0 \times 10^{-13}\text{yr}^{-1} ,$$

and concluded that the presently available data cannot discriminate between the different theories. No evolution correction was considered in the above analysis. Morganstern (1973) has arrived at the same conclusion from a consideration of the observed values of the matter density, the Hubble constant and the deceleration parameter, the ages of different objects in the universe, and the bounds on the variation of G.

We may note one interesting possibility regarding the so called 'missing matter' - in principle this can be provided by the energy of the scalar field. However this would require $(\dot{G}/G) \approx \dot{R}/R \approx 5\times10^{-11}\text{yr}^{-1}$ which is consistent with the Van Flandern (1974) result. But then the presently believed low values of q_0 would be a big question mark.

9.6. ANISOTROPIC SOLUTIONS IN BRANS-DICKE COSMOLOGY

Some anisotropic homogeneous solutions of the Brans-Dicke equations have been investigated (Matzner, Ryan, and Toton 1973; Nariai 1972) especially in connection with the mode of approach to the singularity. However some of the results of Matzner *et al.* are not in agreement with those of Nariai. Further, Nariai has claimed that with $p = \rho$, the Bianchi type I group cosmologies admit non-singular solutions in which the density is always finite. However this result is incorrect as may be easily seen.

From equations (9.7) and (9.8) we have with $v^{\mu} = \delta^{\mu}_{0}$ (cf. the Raychaudhuri equation)

$$\theta_{,\alpha}v^{\alpha} + \frac{1}{3}\theta^{2} - \dot{v}^{\alpha}_{;\alpha} + 2(\sigma^{2}-\omega^{2}) + 4\pi G(\rho+3p+4\rho_{\lambda}) = 0 . \qquad (9.29)$$

Again the divergence of equation (9.7) gives

$$(p+\rho)\dot{v}^{\mu} + [(p+\rho)v^{\nu}]_{;\nu}\, v^{\mu} - p_{,\nu}\, g^{\mu\nu} + \frac{T}{2}\frac{\lambda_{\nu}}{\lambda}\, g^{\mu\nu} = 0 . \qquad (9.30)$$

Contracting with v_{μ}

$$[(p+\rho)v^{\nu}]_{;\nu} - p_{,\nu}\, v^{\nu} + \frac{T}{2}\frac{\lambda_{\gamma}}{\lambda}\, v^{\nu} = 0 . \qquad (9.31)$$

Now for a non-rotating, homogeneous universe we have

$$p_{,\nu} = \alpha v_{\nu},\ \lambda_{\nu} = \beta v_{\nu},\ \omega = 0\ ,\ \dot{v}^{\mu} = 0 . \qquad (9.32)$$

Equation (9.29) therefore gives

$$\theta_{,\alpha}v^{\alpha} + \frac{1}{3}\theta^{2} + 2\sigma^{2} + 4\pi(\rho+3p+4\rho_{\lambda}) = 0 . \qquad (9.33)$$

Equation (9.33) leads to the conclusion that the universe must have a singular state where the expansion scalar is infinite.

The result that ρ remains finite is however interesting and justifies an investigation of the Bianchi type I universe in Brans-Dicke theory (Raychaudhuri 1975a; Ruban and Finkelstein 1975, 1976). The field equations with the line element

$$ds^{2} = dt^{2} - e^{2\phi}dx^{2} - e^{2\psi}dy^{2} - e^{2\theta}dz^{2} , \qquad (9.34)$$

(ϕ, ψ, and θ are functions of t alone) are

$$8\pi G(p+\rho_{\lambda}) = -\ (\ddot{\psi}+\ddot{\theta}) + \frac{3\dot{R}}{2R}\,(\dot{\phi}-\dot{\psi}-\dot{\theta})$$

$$-\ \frac{1}{2}\,(\dot{\phi}^{2}+\dot{\psi}^{2}+\dot{\theta}^{2})$$

$$= -(\ddot{\phi}+\ddot{\theta}) + \frac{3\dot{R}}{2R}(\dot{\psi}-\dot{\theta}-\dot{\phi})$$

$$- \frac{1}{2}(\dot{\phi}^2+\dot{\psi}^2+\dot{\theta}^2)$$

$$= -(\ddot{\psi}+\ddot{\phi}) + \frac{3\dot{R}}{2R}(\dot{\theta}-\dot{\phi}-\dot{\psi})$$

$$- \frac{1}{2}(\dot{\phi}^2+\dot{\psi}^2+\dot{\theta}^2)$$

$$8\pi G(\rho+\rho_\lambda) = \frac{9}{2}\frac{\dot{R}^2}{R^2} - \frac{1}{2}(\dot{\phi}^2+\dot{\psi}^2+\dot{\theta}^2) , \tag{9.35}$$

where $R^3 \equiv \exp(\theta+\phi+\psi)$ and

$$p_\lambda = \rho_\lambda = \frac{2\omega+3}{32\pi G}\left(\frac{\dot{\lambda}}{\lambda}\right)^2 \tag{9.36}$$

From equations (9.34) and (9.35) with $p = \rho$, we get

$$\frac{8\pi G}{3}(\rho+\rho_\lambda)R^6 = a , \tag{9.37}$$

where a is a positive constant. Again the divergence of equation (9.7) yields in this case

$$\frac{8\pi G}{3}\rho R^6\lambda^{-1} = b \tag{9.38}$$

where b is another positive constant. From equation (9.34) we have

$$\begin{aligned} \dot{\phi} &= \alpha R^{-3} + R^{-1}\dot{R} \\ \dot{\psi} &= \beta R^{-3} + R^{-1}\dot{R} \\ \dot{\theta} &= \gamma R^{-3} + R^{-1}\dot{R} \end{aligned} \tag{9.39}$$

where α, β, γ are constants with the constraint

$$\alpha+\beta+\gamma = 0 . \tag{9.40}$$

Substituting from equations (9.37) and (9.39) in (9.35) we get

$$\dot{R}^2 R^4 = a + \tfrac{1}{2}(\alpha^2+\beta^2+\gamma^2) \equiv A^2 \qquad (9.41)$$

where $A^2 \geq a$, the sign of equality corresponding to vanishing shear (isotropic case). Equation (9.41) integrates to give

$$R^3 = 3At , \qquad (9.42)$$

where the origin of time has been chosen to coincide with the epoch $R = 0$. Eliminating ρ and ρ_λ from equations (9.36) - (9.38), we have

$$b\lambda + \frac{2\omega+3}{12}\,\frac{R^6}{\lambda^2}\left(\frac{\mathrm{d}\lambda}{\mathrm{d}R}\right)^2 \dot{R}^2 = a .$$

With the help of (9.42) this may be integrated to give

$$\frac{1\mp(1-b\lambda a^{-1})^{\frac{1}{2}}}{1\pm(1-b\lambda a^{-1})^{\frac{1}{2}}} = CR^{\pm(12/2\omega+3)^{\frac{1}{2}}\frac{a^{\frac{1}{2}}}{A}}$$

where C is an arbitrary constant. Using the positive sign in the exponent of R (the conclusions are the same even if we choose the negative exponent), we get near the singularity $R \to 0$,

$$\lambda \propto R^{(12/2\omega+3)^{\frac{1}{2}}\frac{a^{\frac{1}{2}}}{A}} . \qquad (9.43)$$

Equation (9.38) shows that ρ will remain finite as $R \to 0$ if

$$\left(\frac{12}{2\omega+3}\right)^{\frac{1}{2}} \frac{a^{\frac{1}{2}}}{A} = 6$$

or,

$$0 < 2\omega+3 = \frac{A^{-2}a}{3} \leq \frac{1}{3} , \qquad (9.44)$$

the sign of equality corresponding to the case of vanishing shear. Relation (9.44) is essentially the same as the condition given by Nariai.

Thus the singularity, here, consists in the vanishing

of spatial volumes and infinities in the value of the shear, the expansion, and the scalar field energy but the matter energy remains finite. However, this case corresponds to values of ω which would make the Brans-Dicke theory inconsistent with observations and the assumed equation of state $p = \rho$ is also unphysical. Also the state of affairs at the singularity looks different if one translates the results to Dicke's unrevised version. In case $p = \rho$, the results are for $2\omega+3 = 1/3$, $aA^{-2} = 1$,

$$\phi \sim \frac{T}{(T+1)^2} ,$$

$$R \sim T^{-1/3}(T+1) ,$$

$$\rho \sim T^2(T+1)^{-6}$$

$$G \sim (T+1)^2 T^{-1} .$$

Thus the spatial volume never vanishes for $0 \leq T$ and is arbitrarily large for $T \to 0$ or $T \to \infty$. At these limits ρ vanishes and G tends towards infinity. Thus the singularity appears not as a collapse of space, rather there is a finite maximum of ρ at $T = \frac{1}{2}$.

Indeed, if one allows negative values of ω as in the above case, one can have also a class of solutions in which the geometry is static although G changes with time. Thus with the Robertson-Walker line element if R be a constant, one has the field equations

$$\ddot{\phi} = \frac{8\pi}{(2\omega+3)} (\rho - 3p) ,$$

$$\frac{k}{R^2} = \frac{8\pi\rho}{3\phi} + \frac{\omega\dot{\phi}^2}{6\phi^2} .$$

An obvious solution with $p = \varepsilon\rho$ is

$$\phi = \left(\frac{4\pi\rho}{\omega}\right) t^2 ,$$

$$\omega = -\frac{1}{1-\varepsilon} ,$$

$$k = 0$$

so that

$$G = \left(\frac{2\omega+4}{2\omega+3}\right)\phi^{-1} \sim \left(\frac{2-4\varepsilon}{2-3\varepsilon}\right) \frac{1}{4\pi\rho(1-\varepsilon)} t^{-2} .$$

G is positive for $\varepsilon < \frac{1}{2}$ and has singular behaviour at the limits $t \to 0$ and $t \to \infty$. Thus these solutions, admittedly not of physical interest, nevertheless widen our horizon about the nature of the singularities, which as we now go to show, must necessarily occur in Brans-Dicke theory for $2\omega+3 \geq 0$.

9.7. SINGULARITY IN SCALAR TENSOR THEORIES

We begin with a reference to the Hawking-Penrose theorem. For a normal geodesic congruence v^a, the Raychaudhuri equation reads

$$-\frac{d\theta}{dS} = 2\sigma^2 + \frac{1}{3}\theta^2 - R_{ab}v^a v^b .$$

The Hawking-Penrose theorem on the existence of singularities follow if the r.h.s. is always positive for any time-like v^a so that the geodesics show a monotone convergence due to gravitational attraction. Thus one arrives at the condition that for any time-like v^a one must have $R_{ab} v^a v^b \leq 0$ or with

$$R_{\mu\nu} = -8\pi(T_{\mu\nu} - \tfrac{1}{2} Tg_{\mu\nu}) ,$$

$$(T_{\mu\nu} - \tfrac{1}{2} Tg_{\mu\nu}) v^\mu v^\nu \geq 0 .$$

We have already seen that in the Dicke revised version, the general relativity equation holds provided one takes for

$T_{\mu\nu}$ the sum of that due to known fields and the scalar field of Brans and Dicke. Thus the Hawking-Penrose condition will be satisfied if

$$(\Lambda_{ij} - \tfrac{1}{2}\Lambda g_{ij})\, v^i v^j \geq 0 \; .$$

Using (9.8) we have

$$(\Lambda_{ij} - \tfrac{1}{2}\Lambda g_{ij})\, v^i v^j = \frac{2\omega+3}{16\pi G\lambda^2}\,(\lambda_i v^i)^2 \; .$$

Thus the inequality is satisfied if $(2\omega+3) \geq 0$.

The Brans-Dicke scalar field is a massless scalar field as in the Lagrangian (9.14) we have no term of the form $m^2\phi^2$. If such a term were present, one would be led to an energy-momentum tensor of the form

$$T_{\alpha\beta} = \phi_\alpha\phi_\beta - \tfrac{1}{2}g_{\alpha\beta}(\phi_\alpha\phi^\alpha + m^2\phi^2)$$

and consequently $(T_{\alpha\beta}v^\alpha v^\beta - \tfrac{1}{2}T)$ would be equal to

$$(\phi_\alpha v^\alpha)^2 - \tfrac{1}{2}m^2\phi^2$$

which may not satisfy the inequality. The difficulty may indeed occur for pion fields. However this can only affect the inequality over distances not greater than 10^{-12} cm. (Hawking and Ellis 1973). Bekenstein (1975) has obtained non-singular models in which the matter in the form of dust interacts with a conformal scalar field so that the energy condition is not satisfied.

The C field of Hoyle and Narlikar which leads to the singularity-free steady state cosmological solutions is a negative-energy scalar field. Recently Narlikar (1973, 1974) has suggested that if the divergence of the C field is of the form of a Dirac δ-function, then one could have a singularity free solution in which all the particles are created at a particular epoch (a singular event?). However, after a short instant, the situation is in no way different from big-bang models.

10
COSMOLOGICAL MODELS BASED ON EINSTEIN-CARTAN THEORY

10.1. THE FIELD EQUATIONS OF THE EINSTEIN-CARTAN THEORY

The theory has a fairly long history. It begins with Cartan (1922) who considered geometries with non-symmetric affinities and thought of a possible connection between the intrinsic spin of matter and the antisymmetric part of the affinities.

Some forty years later, the theory made its appearance as a local gauge theory for the Poincare group in space time (Sciama 1962, 1964a,b; Kibble 1961) and in the seventies the ideas developed into a fully fledged theory due mainly to Trautman and his school and to Hehl and his collaborators (Trautman 1972 a,b,c; 1973a, 1975; Kuchowicz 1973; Hehl 1966, 1973, 1974; Hehl and von der Heyde 1973). Without going into the important arguments advanced in favour of the theory (see e.g. Hayashi and Shirafuji 1977) we shall simply be interested in its application to cosmology and especially the possibility of bounce in this theory.

The basic difference of the Einstein-Cartan theory with the general theory of relativity is that the affinities are no longer restricted to be symmetric; and the antisymmetric part of the affinities is coupled with the intrinsic spins of the particles in the field. The geometry is thus no longer Riemannian and both mass and spin (which must be specified to select a definite irreducible representation of the Poincare group) are now linked up with geometry. In the general theory of relativity, one may follow the procedure of Palatini of varying the metric tensor components and the affinities independently and then with the assumption of the symmetry of the affinities, one gets the relation $g_{ik;l} = 0$ as well as the identity of the affinities with the Christoffel symbols (Pauli 1958). Here a similar procedure is adopted and the relation $g_{ik;l} = 0$ is introduced as an assumption. (See however Safko, Tsmaparlis , and Elston 1977.)

We write Γ^i_{kl} for the non-symmetric affinities and introduce the torsion $Q^i_{kl} = \frac{1}{2}(\Gamma^i_{kl} - \Gamma^i_{lk})$. Unlike the affinities, the torsion components form a tensor of rank three. We introduce the contorsion tensor

$$K^{k}_{ij} \equiv -\Gamma^{k}_{ij} + \left\{ {k \atop ij} \right\} , \tag{10.1}$$

where $\left\{ {k \atop ij} \right\}$s are the Christoffel symbols belonging to the metric g_{ik}. It is to be noted that in general K^{k}_{ij} is neither symmetric nor antisymmetric. We shall assume that $g_{ik;l} = 0$, the semicolon indicating covariant derivatives with respect to the Γ^{i}_{kl} s. We then get

$$K^{k}_{ij} = -Q^{k}_{ij} + g^{kl}\, g_{im}\, Q^{m}_{jl} - g^{kl}\, g_{jm}\, Q^{m}_{li} . \tag{10.2}$$

The Riemann-Christoffel tensor is defined by parallel transport along the boundary of a surface element:

$$R^{l}_{ijk} \equiv \Gamma^{l}_{ik,j} - \Gamma^{l}_{jk,i} + \Gamma^{l}_{jm}\,\Gamma^{m}_{ik} - \Gamma^{l}_{im}\,\Gamma^{m}_{jk} . \tag{10.3}$$

The tensor has the usual symmetry properties as in Riemannian geometry except that the cyclic identities are replaced by (Kuchowicz 1973)

$$R^{l}_{[ijk]} = 2Q^{l}_{[kj;i]} - 4Q^{m}_{[ji}Q^{l}_{k]m} \tag{10.4}$$

$$2\, R_{i[jkl]} = \sum_{+3-1} \left\{ 2Q^{l}_{[kj;i]} - 4Q^{m}_{[ji}Q^{l}_{k]m} \right\} \tag{10.5}$$

where in the last equation we have to sum the expression three times with the plus sign for the permutations (ijkl), (jkli), and (klij) and once for (lijk) with the minus sign.

We define

$$R_{ij} \equiv R_{kij}{}^{k}; \quad G_{ij} \equiv R_{ij} - \tfrac{1}{2}g_{ij}R ,$$

and introduce the notation

$$A^{l\ldots}_{m*;i} \equiv A^{l\ldots}_{m;i} + 2Q^{k}_{ik}\, A^{l\ldots}_{m} . \tag{10.6}$$

Now the variational principle

$$\delta \int \left[R + \frac{16\pi G}{c^4} L\right] \sqrt{-g}\ dx^0 dx^1 dx^2 dx^3 = 0 \tag{10.7}$$

(where L is the Lagrangian due to non-gravitational fields) yields (by varying the metric tensor g_{ik} and the torsion Q^1_{ik} independently)

$$T^{ij} \equiv \frac{2}{\sqrt{-g}} \frac{\partial}{\partial g_{ij}} (\sqrt{-g}L) = -\frac{1}{8\pi G} [G^{ij} + C^{ijk}{}_{*;k}] \tag{10.8}$$

$$\frac{1}{\sqrt{-g}} \frac{\partial L}{\partial Q_{ij}{}^k} = -2(C_K{}^{ij} - C_K + C^j{}_K{}^i) \tag{10.9}$$

where the tensor C^k_{ij} is defined by

$$C_{ij}{}^k \equiv Q_{ij}{}^k + \delta^k_i Q_{j1}{}^1 - \delta^k_j Q_{i1}{}^1 \tag{10.10}$$

and is antisymmetric in i,j. The field equations become

$$-8\pi G T^{ij} = G^{ij} + C^{ijk}{}_{*;k} \tag{10.11}$$

$$C^{ijk} = 8\pi G S^{ijk} \tag{10.12}$$

where

$$\sqrt{-g}\ S_k{}^{ij} \equiv \frac{\partial L}{\partial K_{ij}{}^k}\ . \tag{10.13}$$

$S_k{}^{ij}$ is the canonically defined spin tensor. We may note that while T^{ij} is by definition symmetric, G^{ij} is non-symmetric in the presence of spin, its antisymmetric part being cancelled by $C^{ijk}{}_{*;k}$ in equation (10.11). The tensor G^{ij} can be broken up into the Riemannian part $(G^{ij})_{\text{Riem}}$ corresponding to $\left\{ {i \atop k1} \right\}$ and that arising from the presence of torsion.

$$G^{ij} = (G^{ij})_{\text{Riem}} - (C^{ijk} - C^{jki} + C^{kij})_{*;k} - 4C^{ik}{}_{[1}C^{j1}{}_{k]}$$
$$-2C^{ik1}C^j{}_{k1} + C^{k1i}C_{k1}{}^j + \tfrac{1}{2}g^{ij}(4C^k{}_{m[1}C^{m1}{}_{k]} + C^{mk1}C_{mk1})\ . \tag{10.14}$$

The Lagrangian L_{EC} of the Einstein-Cartan theory differs from L_{GR} of general relativity by the terms

$$L_{EC}-L_{GR} = 8\pi G(-\tfrac{1}{2}\, S_{ijk}S^{ijk} + S_{ijk}S^{jki} + S_{ik}{}^{k}S^{i}{}_{l}{}^{l}) + \text{a divergence term.} \tag{10.15}$$

This may be regarded as representing a universal spin-spin contact interaction. It is however an extremely weak interaction as evident from the presence of G as a coefficient.

As in the general theory of relativity, the Bianchi identities lead to the conservation principles of energy momentum as also angular momentum (due to intrinsic spin)

$$t^{j}_{i\overset{+}{;}j} = S_{jk}{}^{l}\, R_{il}{}^{jk} \tag{10.16}$$

$$S_{ij}{}^{k}{}_{\overset{*}{;}k} = t_{[ij]}\,, \tag{10.17}$$

where t_{ij} is the canonical (non-symmetric) energy-momentum tensor and the field equations may be alternatively written in the form

$$G^{ij} = -8\pi t^{ij} \tag{10.18}$$

so that

$$t^{ij} = T^{ij} + S^{ijk}{}_{*;k}\,. \tag{10.19}$$

The symbol $\overset{+}{;}$ in equation (10.16) is defined as

$$\psi^{\cdots\cdot}_{j\;\overset{+}{;}i} = \psi^{\cdots\cdot}_{j\;,i} + 4Q^{k}_{i(j}\,\psi^{\cdots\cdot}_{k)} \tag{10.20}$$

10.2. COSMOLOGICAL MODELS OF THE EINSTEIN-CARTAN THEORY

Let us take the simplest case, the Bianchi type I models, first of all (Kopczynski 1972, 1973; Trautman 1973b; Raychaudhuri 1975b). The line element is

$$ds^2 = dt^2 - e^{2\phi}dx^2 - e^{2\psi}dy^2 - e^{2\theta}dz^2 \tag{10.21}$$

with ϕ, ψ, and θ functions of t alone. The Ricci tensor components are with the Christoffel symbols as connections

$$(G^1_1)_{\text{Riem}} = -(\ddot{\psi}+\ddot{\theta}) + (\dot{\phi}+\dot{\psi}+\dot{\theta})\left(\frac{\dot{\phi}}{2} - \frac{\dot{\psi}}{2} - \frac{\dot{\theta}}{2}\right) - \tfrac{1}{2}(\dot{\phi}^2+\dot{\psi}^2+\dot{\theta}^2), \quad (10.22a)$$

$$(G^2_2)_{\text{Riem}} = -(\ddot{\phi}+\ddot{\theta}) + (\dot{\phi}+\dot{\psi}+\dot{\theta})\left(\frac{\dot{\psi}}{2} - \frac{\dot{\phi}}{2} - \frac{\dot{\theta}}{2}\right) - \tfrac{1}{2}(\dot{\phi}^2+\dot{\psi}^2+\dot{\theta}^2), \quad (10.22b)$$

$$(G^3_3)_{\text{Riem}} = -(\ddot{\psi}+\ddot{\phi}) + (\dot{\phi}+\dot{\psi}+\dot{\theta})\left(\frac{\dot{\theta}}{2} - \frac{\dot{\phi}}{2} - \frac{\dot{\psi}}{2}\right) - \tfrac{1}{2}(\dot{\phi}^2+\dot{\psi}^2+\dot{\theta}^2), \quad (10.22c)$$

$$(G^0_0)_{\text{Riem}} = \tfrac{1}{2}(\dot{\phi}^2+\dot{\psi}^2+\dot{\theta}^2) - \tfrac{1}{2}(\dot{\phi}+\dot{\psi}+\dot{\theta})^2 \quad . \quad (10.22d)$$

The co-ordinate system is co-moving, and we note the following: we introduce the assumption that both the momentum density and the spin density are transported with the fluid velocity v^i (the hydrodynamic or convective description; Kerlick 1976) so that

$$T_{ik} = (p+\rho)v_i v_k - pg_{ik} \quad (10.23a)$$

$$S_{ij}{}^k = S_{ij}v^k \quad . \quad (10.23b)$$

The spin tensor S_{ij} has six independent components, but as the spin tensor should have only three independent components (Papapetrou 1951) we adopt the assumption used originally by Weyssenhoff and Raabe (1947) and subsequently by others in Einstein-Cartan models:

$$S_{ij}v^j = 0 \quad . \quad (10.24)$$

We suppose that the spins of all the particles have been oriented along the x-axis so that the only non-vanishing component of S_{ij} is $S_{23} = -S_{32}$. How this alignment of the spins may be brought about remains somewhat obscure in the theory and we shall return to this question later on. With the specifications (10.23) and (10.24), (10.12) and (10.10) yield

$$8\pi S_{ij}{}^k = C_{ij}{}^k = Q_{ij}{}^k , \quad Q_{i1}{}^1 = 0 \quad (10.25)$$

so that the $\overset{*}{;}$ derivatives reduce to the covariant derivative (see 10.6). Equation (10.2) now give for the non-vanishing k^{1}_{ik} with the assumed spin alignment

$$\begin{aligned} K_{23}{}^{0} &= -K_{32}{}^{0} = -Q_{23}{}^{0} = -8\pi S_{23}\,, \\ K_{02}{}^{3} &= K_{20}{}^{3} = -e^{-2\theta}Q_{23}{}^{0} = -8\pi e^{-2\theta}S_{23}\,, \\ K_{03}{}^{2} &= K_{30}{}^{2} = +e^{-2\psi}Q_{23}{}^{0} = 8\pi e^{-2\psi}S_{23}\,. \end{aligned} \tag{10.26}$$

The non-vanishing components of $C^{ijk}_{\;\;;k}$ are

$$\left.\begin{aligned} -C^{32K}{}_{;k} &= C^{23K}{}_{;k} = 8\pi[S^{23}_{,0} + S^{23}(\dot\phi+2\dot\psi+2\dot\theta)] \\ -C^{2K3}{}_{;k} &= C^{K23}{}_{;k} = 8\pi\dot\theta S^{23} \\ -C^{3K2}{}_{;k} &= C^{K32}{}_{;k} = 8\pi\dot\psi S^{23} \\ -C^{0K0}{}_{;k} &= C^{K00}{}_{;k} = 128\,\pi^{2}S_{23}S^{23} \\ -C^{2K2}{}_{;k} &= C^{k22}{}_{;k} = 64\,\pi^{2}S_{23}S^{23} \\ -C^{3K3}{}_{;k} &= C^{K33}{}_{;k} = 64\,\pi^{2}S_{23}S^{23}\,. \end{aligned}\right\} \tag{10.27}$$

Using equations (10.27) in (10.14) and (10.11) we get $T^{10} = T^{20} = T^{30} = T^{23} = T^{13} = T^{12} = 0$, and

$$\left.\begin{aligned} -8\pi T^{1}_{1} &= +(G^{1}_{1})_{\text{Riem}} + 64\pi^{2}\, S_{23}S^{23} \\ -8\pi T^{2}_{2} &= (G^{2}_{2})_{\text{Riem}} + 64\pi^{2}\, S_{23}S^{23} \\ -8\pi T^{3}_{3} &= (G^{3}_{3})_{\text{Riem}} + 64\pi^{2}\, S_{23}S^{23} \\ -8\pi T^{2}_{3} &= -8\pi T^{3}_{2}(e^{2\theta}.e^{-2\psi}) \\ &= -e^{2\theta}(\dot\psi-\dot\theta)S^{32} \\ -8\pi T^{0}_{0} &= (G^{0}_{0})_{\text{Riem}} - 64\pi^{2}\, S_{23}S^{23}\,. \end{aligned}\right\} \tag{10.28}$$

Now introducing (10.23) with $p = 0$, we have $T^0_0 = \rho$ and all other components of T^μ_ν vanish, so that

$$\dot{\psi} = \dot{\theta} \tag{10.29}$$

$$\dot{\phi} - \dot{\psi} = 3\alpha R^{-3} , \tag{10.30}$$

where we have written $R^3 \equiv \exp(\phi+\psi+\theta)$ and α is a constant related to the shear. Also from the Bianchi identities one obtains in this case

$$\frac{\partial}{\partial t}(\rho R^3) = 16\pi S \frac{\partial}{\partial t} (SR^3) \tag{10.31}$$

where $S^2 = S_{23}S^{23}$. One can write the solution of the above equation in the form

$$\begin{aligned} \rho &= \rho_1 V^{-1} + \rho_2 V^{-\alpha} \\ S^2 &= S_1^2 V^{-2} + S_2^2 V^{-\alpha} , \end{aligned} \tag{10.32}$$

with $V = R^3$, $\rho_2 = 8\pi S_2^2 \left(\frac{2-\alpha}{1-\alpha}\right)$, ρ_1, ρ_2, S_1, and S_2 are constants. This of course is neither a unique nor the general solution of equation (10.31) but is in a sense typical. The second parts of ρ and S^2 are coupled while the first parts are independently conserved. If we assume that the alignment of the particle spins is frozen, then the first parts alone will be present. Kuchowicz (1975a) has considered a case where both parts are present and found that even then bouncing solutions may be obtained. However if only the coupled second terms are present, then the singularity inevitably occurs (Raychaudhuri and Banerji 1977).

Assuming that the alignment of spins remain frozen, we therefore have

$$\rho R^3 = \frac{3}{4\pi} M \tag{10.33}$$

$$S^2 R^6 = \text{const.} = \Sigma^2 . \tag{10.34}$$

Equations (10.28) give, using (10.29), (10.30), (10.33), and

(10.34)

$$\frac{6M}{R^3} = \frac{3\dot{R}^2}{R^2} + \frac{\mu}{R^6} \tag{10.36}$$

with

$$\mu = 64\pi^2 \Sigma^2 - 3\alpha^2 .$$

Equation (10.36) shows that R will have a minimum if μ be positive and the minimum occurs at $R^3_{min} = \mu/6M$. In the shearfree case considered by Trautman (1973b), $M = Nm$ and

$$\mu = 64\pi^2\Sigma^2 = 64\pi^2\left(\frac{N\hbar}{2}\right)^2 \frac{1}{(4\pi/3)^2} ,$$

assuming that the universe is populated by particles of mass m and spin $\hbar/2$ and their total number in a volume $4\pi R^3/3$ is N. We thus get finally

$$R^3_{min} = \frac{\mu}{6M} = \frac{3}{2}\,\frac{N\hbar^2}{M} .$$

If we consider a region out to the Hubble length $cH_0^{-1} \sim 10^{28}$ cm, and the present number density of baryons to be $\sim 10^{-5}$ cm^{-3}, $N \sim 10^{80}$, and using for m the neutron mass $1{\cdot}7 \times 10^{-24}$g, R_{min} 1 cm and the corresponding density $\rho_{max} \sim 10^{55}$ g cm^{-3}. The integral of equation (10.36) is

$$R^3 = \frac{9Mt^2}{2} + \frac{\mu}{6M}$$

The minimum would be pushed to even higher values of the density if the shear be present and there would be a singularity if $3\alpha^2 \geq 64\pi^2\Sigma^2$. This is of a very small magnitude and is by no means outside the presently observed limits. For $t^2 \gg \mu/(27M^2)$, the behaviour of the model would be indistinguishable from the relativity model. The critical time comes out as $\sim 10^{-25}$ second, so that the subsequent thermal history as well as the nucleosynthesis would proceed as in the standard relativity models.

The bounce depends also on the alignment of the spins. It has been argued by Hehl, Von der Heyde, and Kerlick (1974) that even in the absence of spin alignment, the spin-spin interaction terms in the field equations would not vanish - observe for example the last terms in (10.14) and (10.15). However, in such cases, presumably the bounce will occur at even higher densities. One may imagine with Trautman that any cosmic magnetic field that may be present would increase inversely as the square of linear dimensions, i.e. as R^{-2} if the flux be conserved. (This is, however, not quite correct, for a magnetic field, as we shall presently see, is necessarily associated with shear. Hence the magnetic field varying inversely as the area of elements orthogonal to the magnetic field may have a time dependence quite different from R^{-2}.) Again if one considers an ultrarelativistic gas regime, the temperature increases as R^{-1}. One may thus expect that the Boltzmann factor $\exp[-\mu H(1-\cos\theta)]/kT$ determining the probability of orientation of elementary magnets of moment μ at an angle θ to the magnetic field would be vanishingly small for $\theta \neq 0$.

It is fairly easy to deduce some relations which bring out clearly the effect of spin in the Einstein-Cartan theory. In view of equations (10.24) and (10.25) we have the divergence $A^{\nu}_{;\nu} = A^{\nu}_{;\nu} + \{^{\nu}_{\alpha\nu}\}A^{\alpha} = A^{\nu}_{|\nu}$ where the vertical bar indicates covariant derivative with the Christoffel symbols and $v^{\mu}_{;\nu}\, v^{\nu} = v^{\mu}_{|\nu}\, v^{\nu}$ where v^{i} is the fluid-velocity vector. Also $v_{(\alpha;\beta)} = v_{(\alpha|\beta)}$ so that the expansion, acceleration, and shear are not directly changes in going from the symmetric to the non-symmetric affinities. However for the vorticity tensor

$$[\omega_{\alpha\beta}]_{E.C.} \equiv \tfrac{1}{2}(v_{\alpha;\beta}-v_{\beta;\alpha}) + \tfrac{1}{2}(\dot{v}_{\beta}v_{\alpha}-\dot{v}_{\alpha}v_{\beta})$$

$$= \tfrac{1}{2}(v_{\alpha|\beta}-v_{\beta|\alpha}) + \tfrac{1}{2}(\dot{v}_{\beta}v_{\alpha}-\dot{v}_{\alpha}v_{\beta}) + Q^{\sigma}_{\alpha\beta}v_{\sigma}$$

$$= [\omega_{\alpha\beta}]_{G.R} + 8\pi S_{\alpha\beta} ,$$

and consequently the Raychaudhuri equation now assumes the form (Stewart and Hajicek 1973; Tafel 1973)

$$\theta_{,\alpha}v^{\alpha} + \frac{1}{3}\theta^{2} - \dot{v}^{\alpha}_{;\alpha} + 2\sigma^{2} - (\omega_{ik}+8\pi S_{ik})(\omega^{ik}+8\pi S^{ik}) + 4\pi(\rho+3p) = 0,$$

where the intrinsic spin plays a somewhat similar role to vorticity. Tafel has also given the equations similar to equations (5.16) and (5.17) (p.85):

$$8\pi\rho = \frac{1}{3}\theta^2 - \sigma^2 - \tfrac{1}{2}R^* + 16\pi\omega_{ab}S^{ab} + 64\pi^2 S^2 ,$$

$$\frac{\partial}{\partial t}(R^* + 2\sigma^2 - 32\pi\omega_{ab}S^{ab}) + \frac{2}{3}\theta(R^* + 6\sigma^2 - 4\omega^2 - 64\pi\omega_{ab}S^{ab}) = 0 .$$

Tafel (1975) has presented a solution for the space time admitting Bianchi type V group and again the crucial factor determining whether there will be a bounce or not is the relative magnitude of the spin and shear. For the Bianchi type IX however an investigation by Kerlick (1975a) shows that a homogeneous distribution of polarized spin is inconsistent with spatial closure in this case. (Kuchowicz, 1978).

The field equations of the Einstein-Cartan theory may be recast into a quasi-Einsteinian form with a modified energy momentum tensor $T^*_{\mu\nu}$. Hehl *et al.* (1974) have shown that this tensor may violate the Hawking-Penrose energy condition and thus the existence of singularity-free solutions may be understood.

10.3. MAXWELL EQUATIONS AND COSMOLOGICAL MODELS WITH A MAGNETIC FIELD

The electromagnetic fields do not couple with the torsion (see, however, Novello 1976). To see this in a simple manner consider the variational principle

$$\delta \int [\tfrac{1}{4} F_{\mu\nu}F^{\mu\nu} + J^{\mu}A_{\mu}]\sqrt{-g}\, dx^0 dx^1 dx^2 dx^3 = 0 , \qquad (10.37)$$

where supposing for the moment

$$F_{\mu\nu} \equiv A_{\mu;\nu} - A_{\nu;\mu} = A_{\mu,\nu} - A_{\nu,\mu} - 2Q^{\sigma}_{\mu\nu}A_{\sigma} \neq A_{\mu|\nu} - A_{\nu|\mu} .$$

Thus the gauge invariance is lost. The variational principle gives

$$F^{\mu\nu}_{\ \ ;\nu} = J^{\mu}$$

or,

$$F^{\mu\nu}_{\ |\nu} + Q^{\mu}_{\alpha\nu}F^{\alpha\nu} = J^{\mu} \tag{10.38}$$

where we have used equation (10.24). From equation (10.38) we get

$$\frac{1}{(-g)^{\frac{1}{2}}}(Q_{\alpha}{}^{\mu}F^{\alpha\nu}\sqrt{-g})_{,\mu} = J^{\mu}_{\ ;\mu} = J^{\mu}_{\ |\mu} = \frac{1}{(-g)^{\frac{1}{2}}}(J^{\mu}\sqrt{-g})_{,\mu}$$

or

$$[(J^{\mu}-Q^{\mu}_{\alpha\nu}F^{\alpha\nu})(-g)^{\frac{1}{2}}]_{,\mu} = 0\ . \tag{10.39}$$

Thus if J^{μ} is to be identified as the charge-current vector, then in general the charge-conservation principle does not hold good. It therefore seems appropriate to assume that there is no coupling between the torsion and the electromagnetic field so that the electromagnetic field is to be defined as

$$F_{\mu\nu} = A_{\mu|\nu} - A_{\nu|\mu}$$

when we shall have the usual field equations of Maxwell.

$$F^{\mu\nu}_{\ |\nu} = J^{\mu}$$

$$(J^{\mu}\sqrt{-g})_{,\mu} = 0\ .$$

To have an idea of the influence of magnetic fields in Einstein-Cartan cosmology, let us consider again the line element (10.21) with the energy-momentum tensor as due to the fluid cum electromagnetic field

$$T^{\mu}_{\nu} = (p+\rho)v^{\mu}v_{\nu}-p\delta^{\mu}_{\nu} + \frac{1}{4\pi}(\tfrac{1}{4}F_{\alpha\beta}F^{\alpha\beta}\delta^{\mu}_{\nu} - F^{\mu\alpha}F_{\nu\alpha})\ . \tag{10.40}$$

In general there will be an interaction between the spins and the magnetic field tending to bring about an alignment of the spins.

However, for simplicity we shall restrict ourselves to the condition when the alignment has been brought about already and so no torque will be exerted on the elementary dipoles. We shall take the spins as well as the magnetic field to be in the x-direction, so that only the components S_{23} and F_{23} of the spin tensor and the electromagnetic field tensor are non-vanishing.

As before, the condition $T^{23} = 0$ gives $\dot{\psi} = \dot{\theta}$ and the other field equations are

$$G(8\pi p - B^2) = G_1^1 + 32\pi^2 s^2 G^2 \tag{10.41}$$

$$G(8\pi p + B^2) = G_2^2 + 32\pi^2 s^2 G^2 \tag{10.42}$$

$$G(-8\pi\rho - B^2) = G_0^0 - 32\pi^2 s^2 G^2 , \tag{10.43}$$

and assuming that the alignment of the spins remain frozen,

$$32\pi^2 s^2 = S^2 R^{-6} \tag{10.44}$$

$$B^2 = b^2 e^{-4\psi} \tag{10.45}$$

$$d(\rho R^3) + 3pR^2 dR = 0 , \tag{10.46}$$

where S and b are constants and, as before, $R^3 = \exp(\phi+\theta+\psi)$. In the above equations B represents the strength of the magnetic field and G^α_β is given by equations (10.22). Equations (10.41) and (10.42) yield

$$\frac{d}{dt}\,[R^3(\dot{\psi}-\dot{\phi})] = 2B^2R^3G . \tag{10.47}$$

Also equations (10.42) and (10.43) together give

$$\frac{d}{dt}\,[R^3(\dot{\psi}+\dot{\phi})] = 8\pi G R^3(\rho - p) . \tag{10.48}$$

To proceed further let us first take $p = 0$, so that from (10.46) and (10.47) we get

$$(\dot{\phi}+\dot{\psi}) = AtR^{-3} , \tag{10.49}$$

where the constant A is equal to $8\pi G\rho R^3$ and an arbitrary constant of integration has been absorbed in the time variable. In view of equation (10.49) and the definition of R, we have

$$\dot{\psi} = 3\dot{R}R^{-1} - AtR^{-3} \tag{10.50}$$

$$\dot{\phi} = -3\dot{R}R^{-1} + 2AtR^{-3} \tag{10.51}$$

Substituting from equations (10.50) and (10.51) in (10.47) and using (10.45), we get

$$\frac{d^2}{dt^2} R^3 = \frac{3}{2} A + Gb^2 e^{(\phi - 2\psi)} . \tag{10.52}$$

Adding equations (10.41) and (10.42) and using equations (10.50) and (10.51) and also remembering that $p = 0$, we get after a little reduction

$$24At\dot{R}R^{-4} - 30R^{-2}\dot{R}^2 - 6R^{-1}\ddot{R} + R^{-6}(2S^2G^2 - 6A^2t^2) + R^{-3}A = 0 . \tag{10.53}$$

An integration of equation (10.53) in terms of simple functions seems not to be possible and instead one may consider three special cases.

(a) $B = b = 0$. This is the case considered by Trautman (1973b) and Kopczynski (1973). Equation (10.52) is readily integrated to give

$$R^3 = \tfrac{3}{4} A(t + \tfrac{2}{3}\alpha A^{-1})^2 + (G^2S^2 - \tfrac{4}{3}\alpha^2) A^{-1} \tag{10.54}$$

where α is an arbitrary constant of integration related to the shear and another constant of integration has been fixed by using (10.53). As we have already seen, the occurrence of singularity is prevented if $G^2S^2 > 4\alpha^2/3$.

(b) $S = 0$. The solution in this case is (Thorne 1967)

$$t = a_0(e^{\psi}+2\beta)(e^{\psi}-\beta)^{\frac{1}{2}}$$
$$R^3 = e^{3\psi}+ 4\beta e^{2\psi} - 8\beta^2 e^{\psi} \tag{10.55}$$

where

$$a_0^2 = \frac{4}{3} A^{-1} = (6\pi G\rho R^3)^{-1}$$
$$\beta = 3Gb^2 A^{-1} . \tag{10.56}$$

In this case there is always a singularity of vanishing R.

(c) Finally we present a particular solution in the neighbourhood of $t = 0$ to exhibit the existence of singularity-free solutions in the presence of magnetic fields. We take $\exp \psi = \exp -\phi = R^3$ and

$$R^6 = G^2 S^2 t^2 c^{-4} + \mu^2 . \tag{10.57}$$

Equations (10.52) and (10.53) are approximately satisfied if

$$A\mu \ll 2S^2 G^2 c^{-4} \tag{10.58}$$

$$b = \mu S G^{\frac{1}{2}} c^{-1} . \tag{10.59}$$

At the minimum of R both $\dot{\psi}$ and $\dot{\phi}$ vanish but while e^{ψ} is a minimum, e^{ϕ} is a maximum and the universe may be said to have a cigar-like form with the long axis being in the common direction of the magnetic field and the spin. After some time due to the increase of R, the term $G^2 S^2 R^{-6}$ becomes small and the situation approaches that represented by the equations (10.55) and (10.56). The expansion is still anisotropic, although both $\dot{\psi}$ and $\dot{\phi}$ are positive. Still later when $\exp \psi \gg \beta$, the solution approaches the isotropic Friedmann universe.

We have

$$R_{\min}^3 = \mu = \frac{bc}{G^{\frac{1}{2}}S} < \frac{2S^2G^2}{Ac^4} = \frac{2S^2G}{8\pi\rho R^3c^4} \sim 1\ \text{cm}^3\ . \qquad (10.60)$$

Thus the minimum occurs at a volume even less than in the corresponding Trautman case. With $R_{\min} < 1$ cm, equation (10.59) gives

$$\frac{GN^2\hbar^2}{4B_{\max}^2C^2} < 1\ ,$$

with $N \sim 10^{79}$, $B_{\max} > 10^{39}$ gauss. The inequality (10.58) and (10.59) set an upper bound to b *vis-à-vis* the spin S and density ρ. However, for a given S, b varies directly as μ $(=R_{\min}^3)$. So as $R_{\min}$ decreases, b decreases, but the decrease of b is accompanied by an increase of B (see equation 10.45). Thus the upper bound of b corresponds to a lower bound of B. The lower bound of B as given above far exceeds the critical value $\sim 10^{13}$ gauss at which one would expect the field to show quantum effects (Euler and Kochel 1935).

An explicit integration of the field equations in terms of simple functions is possible for the rather unphysical case $p = \rho$. Equation (10.48) then gives

$$\dot{\phi}+\dot{\psi} = \alpha R^{-3}$$

so that

$$\dot{\psi} = 3\dot{R}R^{-1} - \alpha R^{-3}$$

$$\dot{\phi} = 2\alpha R^{-3} - 3\dot{R}R^{-1}$$

where α is an arbitrary constant of integration. Equation (10.53) is now replaced by

$$24\alpha\dot{R}R^{-4} - 30R^{-2}\dot{R}^2 - 6R^{-1}\ddot{R} + (2S^2G^2-6\alpha^2-2Q)R^{-6} = 0\ , \qquad (10.61)$$

where the constant Q is defined by

$$Q = 8\pi G\rho R^6\ .$$

Equation (10.61) has the first integral

$$\frac{d}{dt} R^3 = 4\alpha+\beta t R^{-3} \tag{10.62}$$

with

$$\beta \equiv S^2G^2 - 3\alpha^2-Q \ .$$

Thus R^3 would have a minimum if $\beta > 0$ which would correspond in a way to spin dominating over gravitation and shear. The integral of equation (10.62) in this case is

$$\left|R^6-4\alpha R^3t-\beta t^2\right| = A^2 \left[\frac{(\mu-2\alpha)t+R^3}{(\mu+2\alpha)t-R^3}\right]^{2\alpha/\mu}$$

with $\mu^2 = \beta+4\alpha^2$. A is an arbitrary constant giving the value of R^6 at $t = 0$. The minimum occurs at $t = -4\alpha R^3_{\min}\ \beta^{-1}$ with

$$R^3_{\min} = A\left(\frac{\mu-2\alpha}{\mu+2\alpha}\right)^{2\alpha/\mu} \ .$$

In the case of $\beta \leq 0$, the integral of equation (10.62) is

$$|R^6-4\alpha R^3t-\beta t^2| = A^2\exp[4\alpha t(R^3-2\alpha t)^{-1}] \quad \text{for } \beta+4\alpha^2 = 0,$$

$$= A\ \exp[-4\alpha\mu^{-1}\tan^{-1} \frac{R^3-2\alpha t}{\mu t}$$

$$\text{for } -\mu^2 = \beta+4\alpha^2 < 0.$$

The magnetic field and the variables ψ and ϕ are given by

$$B^2 = \beta R^{-12}(R^6-4\alpha R^3t-\beta t^2)$$

$$e^{\psi} = R^3|(R^6-4\alpha R^3t-\beta t^2)|^{-\frac{1}{4}}$$

$$e^{\phi} = R^{-3}|(R^6-4\alpha R^3t-\beta t^2)|^{\frac{1}{2}} \ .$$

If however B vanishes, one has for $\beta \neq 0$, (if $\beta = 0$, one has the case where R_{min} coincides with the singular state $R = 0$).

$$R^3 = (2\alpha\pm\mu)t$$

with

$$\mu^2 = 4\alpha^2+\beta \geq 0.$$

The asymptotic nature of these solutions near $R = 0$ in case $S = 0$ has been studied by Thorne (1967).

Unlike the case of a non-radiative magnetic field one can consider the presence of radiation. Indeed if one considers the microwave background to be primeval, the minimum is pushed down to 10^{-16} cm in place of the Trautman value of 1 cm for dust. (Hehl *et al.* 1976). Again Kerlick (1975b) has found that for a system of Dirac particles the theory leads to a spin-spin interaction which helps rather than opposes the development of a singularity. For all these reasons the potentiality of the Einstein-Cartan theory in giving us an acceptable cosmological model free of singularities, seems rather poor.

11
THE COSMOLOGICAL SINGULARITY IN TWO RECENT THEORIES

11.1. THE TWO TENSOR THEORY OF GRAVITATION

The basic idea of this theory due to Isham, Salam and Strathdee (1971, 1973) (see also Salam 1975) is that if gravitons are particles of spin half and parity positive then one may expect a mixing of the graviton states with other 2^+ quantum states. It is now assumed that there indeed exist transition-matrix elements between the field $f_{\mu\nu}$ representing f^0 mesons and the $g_{\mu\nu}$ field of gravitons. It is postulated that the leptons interact with the $g_{\mu\nu}$ field with the coupling parameter G (the Newtonian gravitational constant) and the f^0 field interacts with hadrons alone with the coupling parameter $G_{\rm f}$ ($G_{\rm f} \gg G$, $G_{\rm f}/G \sim 10^{38}$). The f^0 meson exchange gives rise to strong nuclear force between hadrons. The conventional gravitation between hadrons arises from f-g mixing. Thus while the lepton-lepton gravitational potential arises from exchange of gravitons, the lepton-hadron, as also the conventional gravitation between hadrons, has its origin in the transformation of f^0 mesons into gravitons.

The Lagrangian is assumed to be of the form

$$L = G_{\rm f}^{-1}(-f)^{\frac{1}{2}}R(f)+G^{-1}(-g)^{\frac{1}{2}}R(g)+L_{\rm fg}+L(f,\text{hadrons})+L(g,\text{leptons}).$$

Here $L_{\rm fg}$ is the interaction Lagrangian between f and g fields and is assumed to be

$$L_{\rm fg} = M_{\rm f}^2G_{\rm f}^{-1}(-f)^{\frac{1}{2}}(f^{\alpha\beta}-g^{\alpha\beta})(f^{\lambda\sigma}-g^{\lambda\sigma})(g_{\alpha\lambda}g_{\beta\sigma}-g_{\alpha\beta}g_{\lambda\sigma}) .$$

Thus, except for the coupling term, both the $f_{\mu\nu}$ and $g_{\mu\nu}$ fields obey the Einstein field equations with $T^{\mu\nu}$ arising from hadrons and leptons respectively.

So far the theory has little to offer in the problem of cosmological collapse to a singularity. However when one introduces the asymmetric affine connection one obtains a spin-spin interaction term but the coupling constant with this term is $G_{\rm f}$ instead of G. Thus the effect of this interaction is enhanced and as $R_{\min} \alpha G^{1/3}$, in the present theory the bounce occurs at $R_{\min} \sim 10^{13}$ cm in place of the Trautman value of ~ 1 cm and the corresponding density is $\sim 10^{17}$ g cm^{-3} which is only slightly higher than nuclear densities and seem much more plausible than the fantastic value 10^{55} g cm^{-3} in the Einstein-Cartan theory. However, explicit solutions of the two-tensor theory for cosmological models seem not to have been studied so far - except, of course, the solutions which may be called trivial (Aichelburg 1973), i.e. the solutions in which $f_{\mu\nu} = g_{\mu\nu}$, when the coupling term falls off and one has simply the Einstein equations. Further shear and/or electromagnetic fields would presumably cause an increase of the density at bounce in this theory as well and quite a small shear may force a collapse to a singularity.

11.2. THE CONFORMALLY INVARIANT THEORY OF HOYLE AND NARLIKAR

The theory is based on the idea of conformal invariance and is Machian in the sense that the mass of any particle arises from an interaction of the particle with all other particles in the universe. Thus besides the metric tensor field there is a mass field generated by the particles (Hoyle and Narlikar 1972; for an excellent recent review, see Narlikar 1978).

The field equations are complicated involving besides the Einstein tensor $R_{ik}-\frac{1}{2}g_{ik}R$ and the energy-stress-momentum tensor T_{ik}, terms arising from the mass function and its derivatives. However, owing to the conformal invariance, if (g_{ik},m) be any solution of the equations, a conformally transformed $g^*_{ik} = g_{ik}\Omega^2$ and a modified mass function $m^* = \Omega^{-1}m$ is also a solution (excluding of course the singular cases $\Omega = 0$ or ∞). Utilizing this freedom, one may choose the conformal function Ω in such a way that the masses become constant (i.e. $\Omega = m$) and in that case the field equations reduce to the general relativity equations. There are however some constraints: the reduction to the Einstein equations is attained only for a many particle universe where the response of the universe cancels out the advanced part and doubles the retarded part of the mass function. It thus turns out that it is not meaningful to talk of an empty universe and this situation is also claimed to be in welcome consistency with Machian ideas. However the rotating expanding solutions of general relativity may not be excluded in this theory. It is also claimed that the attractive nature of gravitation follows directly from the theory.

The Friedmann universe with constant masses is a solution but it can as well be represented by the Minkowski metric to which the Friedmann metric is conformal. In particular for the closed Friedmann universe

$$ds^2 = dt^2 - S^2\left[\frac{dr^2}{1-kr^2} + r^2\, d\theta^2 + r^2 \sin^2\theta\, d\phi^2\right]$$

the conformal function Ω is

$$\Omega = \frac{(1+X^2)^{\frac{1}{2}}(1+Y^2)^{\frac{1}{2}}}{2S}$$

where

$$X = \tan \tfrac{1}{2}\,(T+R)$$

$$Y = \tan \tfrac{1}{2}\,(T*R)$$

$$T = \int_0^t S^{-1}dt$$

$$R = \sin^{-1} r\ .$$

Thus one can consider the universe to be Euclidean and static. But then one must take account of the fact that masses become time dependent. One thus no longer interprets the red shift as due to recession of galaxies but due to the variation of mass the energy levels are correspondingly altered and a simple calculation shows that one gets back the usual red-shift formula.

It has been suggested that the variation of mass effectively means a variation of the gravitational interaction which may be represented by a variation of the gravitational constant $G \alpha \tau^{-2}$ where τ is the time variable in the Minkowski frame. According to Barnothy and Tinsley (1973) such a rapid variation of G leads to 'drastic' discrepancies with observations.

The conformal function has a singularity at $t = 0$ and it is argued that at this stage one cannot go over to the Friedmann metric. Thus the big-bang singularity of relativistic cosmology is attributed to a 'bad' conformal frame. In the Minkowski frame there is no singularity at least in the geometric sense. The vanishing of mass which occurs at this epoch and the abrupt change in the sign of a coupling constant are not considered to be that serious.

Hoyle (1975) has proposed a theory of the microwave background on the basis of the theory - essentially the idea is that near the epoch $t = 0$, owing to the small values of the electronic masses, the scattering cross section of the electrons is greatly enhanced and hence a thermalization of the stellar radiation from the earlier epoch is easily achieved.

Narlikar (1977) has attempted an explanation of the anomalous red shifts from different apparently close objects on the hypothesis that the conformal function may have some sharp spatial variation. That would mean a sharp change of mass with a corresponding marked change of red shift. However such a variation of the conformal function is equivalent to a marked departure from homogeneity over a small region. If one is willing to accept that, then the anomalous red shift can be understood in standard relativistic cosmology as well.

Kembhavi (1976) has shown that other cosmological solutions also can be made singularity free by conformal transformations - however this is hardly surprising for a geodesic of finite length would be stretched infinitely by an infinite conformal transformation.

12

FATE OF PERTURBATIONS OF ISOTROPIC UNIVERSES

12.1. THE BASIC PROBLEM

The study of the fate of perturbations of the isotropic models has a twofold importance. On the one hand it would be justified to take the isotropic models as a fair representation of the actual universe only if the models show reasonable stability. Thus even if the Hubble shift of spectral lines from distant galaxies had not made the Einstein static universe unacceptable, Eddington's demonstration (1933) that Einstein's universe is unstable would have been sufficient indication that it cannot be a valid representation.

The second point of interest is the formation of condensations like the observed galaxies out of a homogeneous distribution. One can, of course, argue that the seeds of these galaxies existed right from the very beginning - the big bang - and then there would not be a problem of formation but simply one of unfolding those initial peculiarities into the present galaxies. However it seems more appealing to try to explain the observed characteristics and the formation of galaxies in terms of perturbations. For this to be possible, a perturbation bringing in non-homogeneity in the density distribution must, at least in some circumstances, grow with time and sufficiently rapidly as the big-bang models have only a finite time-span.

At first sight these two requirements - stability on the one hand and growth of perturbations on the other - may appear to be mutually contradictory. One may make a compromise, a perturbation on a linear scale much larger than the dimensions and average separation of the galaxies should be smoothed out while those of a certain critical dimension should grow giving rise to observed galaxies.

In his preliminary investigations, Tolman (1934a) considered the case of perturbations with spherical symmetry in the Einstein static universe and the Friedmann dust universe. Thus, with the line element

$$ds^2 = -e^{\lambda}dr^2 - e^{\mu}\ d\theta^2 + \sin^2\theta d\phi^2) + dt^2 \ ,$$

suppose that for $t < 0$,

$$e^{\mu} = r^2,\ \dot{\mu} = 0,\ \ddot{\mu} = 0 \ ,$$

and at $t = 0$,

$$e^{\mu} = r^2,\ \dot{\mu} = 0,\ \ddot{\mu} = \ddot{\mu}_0(r) \neq 0 \ .$$

Then the field equations give at $t = 0$ for a dust distribution

$$\frac{\partial^2}{\partial t^2} \log \rho = 4\pi\rho - \Lambda,$$

$$4\pi\rho = \Lambda - \frac{3}{2}\ddot{\mu}_0 - \tfrac{1}{2}\ddot{\mu}_0'(r) \ ,$$

$$\frac{\partial}{\partial t} \log \rho = 0 \ .$$

It follows that in regions where the density exceeds the critical value $\Lambda/4\pi$, the density enhancement will continue to grow. Similar results were found to hold for the pressureless Friedmann universe.

The same problem was investigated by Sen (1935, 1936) with a different initial condition. Instead of assuming the continuity of $\dot{g}_{\mu\nu}$ at $t = 0$, Sen assumed the continuity of the density ρ. Even then any rudimentary tendency towards condensation was found to progress with time in case of vanishing pressure but he could not arrive at any very definite conclusions if the pressure was non-vanishing. If one assumes that, besides spherical symmetry, there is also isotropy of the expansion, even after the perturbation, then Raychaudhuri (1952) showed that while density gradients decrease with time, the ratio between the perturbed and the background density diverges more and more from unity.

A thorough investigation on the fate of perturbations, considering that these may consist in changes in the metric, the density distribution, and the velocity field, was undertaken by Lifshitz (1946) and more recently by Lifshitz and Khalatnikov (1963). It was assumed that the departures from the isotropic-model condition are small enough as to permit the neglect of products and higher powers of perturbation terms. One thus had equations linear in the perturbation. However the discussion involved fairly complicated calculations and Hawking (1966) gave a much simpler discussion by considering the fate of physical quantities like the vorticity, the density perturbations, and the gravitational waves. A fairly comprehensive review of the subject, as developed up to 1966 was given by Harrison (1967). (See also Jones, 1976.)

12.2 THE CASE OF NEWTONIAN COSMOLOGY

We shall in this section consider the problem with the Newtonian law of gravitation and adopt the following assumptions:

(i) The perturbation is small in the sense that we are justified in neglecting all terms non-linear in the perturbation.

(ii) The perturbation does not bring in turbulence.

(iii) There is no interaction other than gravitational and that arising from gradients of an isotropic fluid pressure.

(iv) The energy-stress tensor is that of a perfect fluid so that there is no viscosity or heat conduction effects. We recall the basic equations of Newtonian cosmology as given in Chapter 2:

$$\frac{\ddot{R}}{R} = -\frac{4\pi\rho}{3} - \frac{1}{3}\sigma_{ik}\sigma_{ik} + \frac{1}{3}v_{[i,k]}v_{[i,k]} - \frac{1}{3}\left(\int\frac{dp}{\rho}\right)_{,ii} \quad (12.1)$$

$$\frac{1}{R^2}(R^2\omega_{ik})^{\cdot} + \sigma_{il}\omega_{lk} + \sigma_{lk}\omega_{il} = 0 \quad (12.2)$$

$$(\rho R^3)^{\cdot} = 0 \tag{12.3}$$

$$\frac{3\dot{R}}{R} = v_{i,i} = \theta \tag{12.4}$$

$$\phi_{,ii} = -4\pi G\rho \; . \tag{12.5}$$

For the perturbation problem we take

$$\rho = \rho_0(1+s) \tag{12.6a}$$

$$v = v_0(1+\alpha) \tag{12.6b}$$

$$\phi = \phi_0(1+\beta) \tag{12.6c}$$

where s is called the condensation and the subscript 0 indicates the values for the unperturbed universe - of course these will be functions of t alone. All the three quantities s, α, and β are assumed small. From equations (12.3) and (12.6a) we get

$$\frac{\dot{R}}{R} = \frac{\dot{R}_0}{R_0} - \frac{\dot{s}}{3} \tag{12.7}$$

$$\frac{\ddot{R}}{R} = \frac{\ddot{R}_0}{R_0} - \frac{\ddot{s}}{3} - \frac{2}{3}\dot{s}\,\frac{\dot{R}_0}{R_0} \; . \tag{12.8}$$

As the shear and vorticity bring in terms of higher order of smallness, they fall off at our level of approximation. New equations (12.1) and (12.8) give

$$\frac{\ddot{s}}{s} + \frac{2}{3}\dot{s}\,\frac{\dot{R}_0}{R_0} = \frac{4}{3}\pi\rho_0\, s + \frac{1}{3}\left(\int \frac{dp}{\rho}\right)_{,ii} \; . \tag{12.9}$$

Let us assume a functional relationship between p and ρ; $p = f(\rho)$ and change from x s to the co-moving co-ordinates xR_0^{-1}. Equation (12.9) now reads

$$\ddot{s} + 2\dot{s}\,\dot{R}_0 R_0^{-1} - 4\pi\rho_0 s = f' R_0^{-2}\nabla^2 s \tag{12.10}$$

where ∇^2 is the Laplacian in co-moving co-ordinates.

To investigate the fate of normal modes we write

$$s = \mu(t) \exp(i\underline{kr}) \; . \tag{12.11}$$

Equation (12.10) then gives

$$\ddot{\mu} + 2\dot{\mu}\frac{\dot{R}_0}{R_0} - 4\pi\rho_0\mu + \frac{k^2 C_s^2 \mu}{R_0^2} = 0 \; . \tag{12.12}$$

where C_s, the sound velocity, is equal to $(f')^{\frac{1}{2}}$ For $\mu = \mu_0 \exp i\omega t$ one gets

$$\omega^2 - 2i\omega\frac{\dot{R}_0}{R_0} + 4\pi\rho_0\left(1 - \frac{k^2 C_s^2}{4\pi\rho_0 R_0^2}\right) = 0 \; . \tag{12.13}$$

If the background universe is static, $\dot{R} = 0$ and ω will be real or imaginary according as k is greater or less than $(4\pi\rho_0)^{\frac{1}{2}} R_0 C_s^{-1}$. Thus the fluid distribution will be unstable to perturbations of wavelength λ $(= 2\pi R_0/k)$ greater than $\lambda_c = \pi C_s \rho_0^{-\frac{1}{2}}$. This result is usually referred to as the Jeans' condition for instability (Jeans 1929). Equation (12.13) and the Jeans' result was obtained for spherically symmetric perturbations by Bonnor (1957) and obtained quite generally by a number of other investigators (Savedoff and Vila 1962; Hunter 1964).

The Jeans' criterion is obviously not valid for the non-static cosmological background - one can only say that the expansion of the universe ($\dot{R}_0 > 0$) hinders the growth of perturbations and the Jeans' condition is necessary but not sufficient for instability.

Bonnor (1967) has investigated equation (12.13) for the case of closed dust-universes and has found that while the condensations may increase significantly (indeed even by a factor $\sim 10^{13}$) during the life span of the universe, nevertheless this cannot account for the formation of galaxies that occur in the universe. This is because if we assume the original perturbation to be due to thermal fluctuation, then the initial value of μ would be, according to statistical thermodynamics, $\sim N^{-\frac{1}{2}} \sim 10^{-34}$ where N is the number of particles involved (being for a typical galaxy $\sim 10^{68}$). Thus even with an increase of 10^{13} times, the present value of the condensation would be only $\sim 10^{-21}$.

The situation seems more hopeful if one considers the universe to be more or less in a stationary state ($\dot{R} \sim 0$). Again assuming $p = 0$, $c_s = 0$ (or rather we are considering wavelengths long enough to make $c_s^2 \lambda^{-2} \rho_0^{-1} \ll 1$) equation (12.12) leads to an exponential growth

$$\mu = \mu_0 \exp[(4\pi\rho_0)^{\frac{1}{2}} t]$$

and apparently one may have the e-folding time $(4\pi\rho_0)^{-\frac{1}{2}}$ small compared to the age of the universe ($\sim 10^{10}$ years) if $\rho_0 > 10^{-29}$ g cm^{-3}. One might expect such conditions to be realizable in a Lemaitre universe which might have a long coasting time (i.e. $\dot{R} \sim 0$) in the past. We shall return to this point a little later.

We may note that for any perturbation bringing in vorticity, equation (12.2) gives at the level of linear approximation

$$\omega_{ik} R^2 = \text{const.}$$

Thus such perturbations would be decreasing with expansion. We next go over to the relativistic calculations where, as we shall presently see, the formulae are very similar.

12.3. PERTURBATIONS OF THE FRIEDMANN UNIVERSE

In the perturbed universe we take the t-lines as our flow lines and if the co-ordinate lengths along these lines be equal to the proper time intervals, the line element can be written without any loss of generality

$$ds^2 = dt^2 + 2g_{oi}\, dt dx^i + g_{ik}\, dx^i\, dx^k \,. \qquad (12.14)$$

In the above the roman indices run from 1 to 3, while any latin index, introduced hereafter, will run from 0 to 3 (t being considered the zeroth co-ordinate). The velocity vector $v^\mu = \delta_0^\mu$ with this choice of co-ordinates. The smallness of the perturbation will mean that each of the following quantities which vanish for the unperturbed universe is small

$$v_i = g_{oi},\ p,_i;\ \rho,_i;\ g_{ik} + \frac{R_0}{(1+kr^2/4)^2}\delta_{ik}\ .$$

The non-vanishing g_{oi} s indicate the presence of vorticity and/or a translational velocity relative to the frame of isotropy of the unperturbed universe. To our order of approximation

$$g^{00} = 1,$$

$$g^{io} = -g^{ik}\,g_{k0} = R_0^{-2}\left(1 + \frac{kr^2}{4}\right)^2 g_{io}\ . \qquad (12.15)$$

The divergence relation $T^{\mu\nu}_{;\nu} = 0$ gives with the perfect-fluid assumption

$$[(p+\rho)v^{\mu}]_{;\mu}\,v^{\nu} + (p+\rho)\dot{v}^{\nu} - p_{,\mu}g^{\mu\nu} = 0\ . \qquad (12.16)$$

Splitting up the above into components along and orthogonal to v^{μ}, we get

$$\rho_{,\mu}v^{\mu} = -(p+\rho)\theta \qquad (12.17)$$

$$\dot{v}^{\nu} = p_{,\mu}(g^{\mu\nu} - v^{\mu}v^{\nu})/(p+\rho) \qquad (12.18)$$

where the expansion $\theta = v^{\mu}_{;\mu} - 3\dot{R}R^{-1}$; R so defined would differ from R_0 by a small quantity. Equations (12.17) and (12.18) give for the line element (12.14)

$$\dot{\rho} = -3(p+\rho)\dot{R}R^{-1} \qquad (12.19)$$

$$\dot{g}_{oi} = (p_{,i} - pg_{oi})/(p+\rho)\ . \qquad (12.20)$$

The non-vanishing components of the vorticity tensor becomes, again to our order of approximation,

$$\omega_{ik} = \tfrac{1}{2}(g_{0i,k} - g_{0k,i})$$

and hence using equation (12.20)

$$\dot{\omega}_{ik} = - \dot{p}\ \omega_{ik}/(p+\rho) \ . \tag{12.21}$$

For the case of vanishing pressure ω_{ik} s will be constant and hence

$$\omega \equiv (\tfrac{1}{2}\ g^{ik}\ g^{lm}\ \omega_{il}\ \omega_{km})^{\frac{1}{2}}$$

would be given by

$$\omega = \omega_0 R^{-2} \tag{12.22}$$

showing the same dependence on expansion as in the Newtonian case.

For the other extreme case of ultrarelativistic gas and/or radiation $p = \rho/3$ and equatons (12.19) and (12.21) together give

$$\omega = \omega_0 R^{-1} \ . \tag{12.23}$$

ω_0 in the above equations is an arbitrary constant of integration determined by the initial value of the vorticity perturbation. We see that in either case the vorticity dies out with expansion.

To study the fate of density perturbations, we consider the Raychaudhuri equation. The shear and vorticity terms are negligible, being of the second order of smallness. Using equations (12.18) and (12.14) we get

$$\dot{v}^{\alpha}_{;\alpha} = \frac{1}{(p+\rho)}\ [g^{\alpha\beta} p_{,\alpha;\beta} - \ddot{p} - \dot{p}\theta] - \frac{1}{(p+\rho)^2}\ [p_{,}^{\mu}(p+\rho)_{,\mu} - \dot{p}(\dot{p}+\dot{\rho})] \ .$$

Substituting from equation (12.19), the above relation becomes

$$\dot{v}^{\alpha}_{;\alpha} = \frac{1}{p+\rho}\ (\Box^2 p - \ddot{p} - 2\dot{p}\theta) - \frac{p_{,}^{\mu}(p+\rho)_{,\mu} - \dot{p}^2}{(p+\rho)^2} \tag{12.24}$$

Also,

$$\Box^2 p = \nabla^2 p + \ddot{p} + \dot{p}\theta - \dot{p}(-g)^{-\frac{1}{2}}[g^{ik} g_{io}\sqrt{-g}]_{,k} \; ,$$

∇^2 representing the Laplacian for the three-space metric $ds^2 = g_{ik}\, dx^i dx^k$. Hence, finally

$$\dot{v}^{\alpha}_{;\alpha} = \frac{1}{(p+\rho)} [\nabla^2 p - \dot{p}(-g)^{-\frac{1}{2}}(g_{io} g^{ik}\sqrt{-g})_{,k}] \; . \qquad (12.25)$$

Introducing the condensation s, we have from equation (12.19)

$$\frac{\dot{R}_0}{R_0} - \frac{\dot{s}}{3}\frac{\rho_0}{p_0+\rho_0} = \frac{\dot{R}}{R}\left[1 - \frac{p_0 s - (p-p_0)}{(p_0+\rho_0)}\right] \; . \qquad (12.26)$$

To proceed further we assume a linear relation between p and $p = \alpha\rho$. We then have

$$\frac{\dot{R}}{R} = \frac{\dot{R}_0}{R_0} - \frac{\dot{s}}{3(1+\alpha)} \qquad (12.27)$$

$$\frac{\ddot{R}}{R} = \frac{\ddot{R}_0}{R_0} - \frac{\ddot{s}}{3}\frac{1}{(1+\alpha)} - \frac{2\dot{s}}{3(1+\alpha)}\frac{\dot{R}_0}{R_0} \; . \qquad (12.28)$$

Substituting from equations (12.25), (12.27) and (12.28) in the Raychaudhuri equation, we get

$$\ddot{s} + \frac{2\dot{R}_0}{R_0}\dot{s} + \alpha\nabla^2 s + \frac{3\alpha(1+\alpha)}{\sqrt{-g}}\frac{\dot{R}_0}{R_0}(g_{io}g^{ik}\sqrt{-g})_{,k} - 4\pi\rho_0(1+3\alpha) \times$$

$$(1+\alpha)s = 0 \; . \qquad (12.29)$$

Also from equation (12.20)

$$\dot{g}_{oi} = \frac{\alpha s_{,i}}{1+\alpha} + 3\alpha\frac{\dot{R}_0}{R_0} g_{oi} \; ,$$

which on integration gives

$$g_{oi} = \frac{\alpha}{1+\alpha} R_0^{-3\alpha} \int R_0^{3\alpha}\, s_{,i}\, dt + f_i R_0^{-3\alpha} \; , \qquad (12.30)$$

where f_i s are functions of space co-ordinates alone. The first term in g_{0i} is a gradient $\phi_{,i}$ and can be removed by the infinitesimal transformation $t' = t+\phi(x)$. In any case it does not contribute to vorticity but indicates a 'tilting' in the sense of King and Ellis (1973). The second term gives rise to the term

$$\alpha \dot{R}_0 R_0^{-1-3\alpha} g^{ik} f_{i,k}$$

in equation (12.29). This term has a time dependence $t^{-3/2}$ for the radiation universe in which $R_0 \sim t^{\frac{1}{2}}$ and vanishes for the dust case ($\alpha = 0$). Locally $f_{i,k} + f_{k,i}$ may be made to vanish by a suitable transformation and in any case the term decreases with time. We therefore disregard this term in our subsequent discussion.

We are thus led to the equation

$$\ddot{s} + 2R_0^{-1}\dot{R}_0\dot{s} + \alpha\nabla^2 s - 4\pi\rho_0(1+3\alpha)(1+\alpha)s = 0 . \tag{12.31}$$

This well known linear density perturbation relation has been obtained by a number of investigators (Bonnor 1957; Brecher and Silk 1969; Irvine 1965; Bandyopadhyay 1977). The consequences of this equation have been studied extensively (e.g. Liang 1976) and has been integrated in closed form in some special cases (Groth and Peebles 1975; Adams and Canuto, 1975; Olson, 1976).

We write $s = \mu(t)\psi(\underline{r})$, the equation (12.31) splits up into two equations

$$\ddot{\mu} + 2\dot{\mu}\frac{\dot{R}_0}{R_0} - \alpha\frac{K^2\mu}{R_0^2} - 4\pi\rho_0\mu(1+3\alpha)(1+\alpha) = 0 \tag{12.32}$$

$$\nabla^2\psi = -K^2\psi R_0^{-2} . \tag{12.33}$$

Equation (12.33) reads in spherical polar co-ordinates

$$\frac{1}{q^{3/2}r^2}\frac{\partial}{\partial r}\left(q^{\frac{1}{2}}r^2\frac{\partial\psi}{\partial r}\right) + \frac{1}{r^2\sin\theta}\frac{\partial}{\partial\theta}\left(\sin\theta\frac{\partial\psi}{\partial\theta}\right) + \frac{1}{r^2}\frac{\partial^2\psi}{\partial\phi^2} = K^2\psi, \tag{12.34}$$

with $q = (1+kr^2/4)^{-2}$. Solutions of equation (12.34) can be written in the form $\psi = R(r)\, Y_{lm}(\theta,\phi)$ where Y_{lm} s are the usual spherical harmonics. The solutions $R(r)$ of the radial equation may be found fairly easily (Fock 1935; Schrodinger 1939, 1957). The eigenvalues K for $k = 0, -1$ (open-space cases) have all possible values satisfying $K^2 \geq k^2$ while for $k = +1$, the eigenvalues are discrete, $K^2 = \nu(\nu+2)$ where ν has the positive integral values 1,2,3 etc. Over regions r for which $|kr^2| \ll 2$, $q \approx 1$, and hence the radial function $R(r)$ approach the form $\exp(iKr)/r$. Thus for regions small compared to the dimensions of the universe, the perturbations are approximately of the wave form with characteristic dimension $2\pi K^{-1}R_0$, the co-ordinate length being $2\pi K^{-1}$.

Let us now consider the two extreme cases $\alpha = 0$ and $\alpha = 1/3$. In the first case equation (12.32) reduces to equation (12.12) with $C_s = 0$ and we get back the Jeans' criterion of instability as well as other results of Newtonian theory. In the second case we get

$$\ddot{\mu} + 2\dot{\mu}\frac{\dot{R}_0}{R_0} + \frac{K^2}{3R_0^2}\mu - \frac{32\pi\rho_0\mu}{3} = 0 . \qquad (12.35)$$

In the early phase $R_0 = at^2$, $32\pi\rho_0/3 = t^{-2}$, $a^4 = 32\pi\rho_0 R_0^4/3$. We consider two limiting cases of equation (12.35). If $K^2 \ll 32\pi\rho_0 R_0^2$ or $t \ll K^2/3a^2$, equation (12.35) may be integrated to give $\mu = At+Bt^{-1}$, here, although the first term shows a growth of the perturbation, nevertheless the growth is of a limited extent as the solution holds only up to $t \ll (Ka^{-1})^2/3$. At the opposite extreme of $K^2 \gg 32\pi\rho_0 R_0^2$ or $t \gg K^2a^{-2}/3$, equation (12.35) may be written

$$\ddot{\mu} + 2\dot{\mu}\frac{\dot{R}_0}{R_0} + \frac{K^2}{3R_0^2}\mu = 0 , \qquad (12.36)$$

and this has the approximate solution

$$\mu = AR_0^{-\frac{1}{2}}\, e^{i(t/3)^{\frac{1}{2}}K} \qquad (12.37)$$

representing oscillation with decreasing amplitude. Thus if the 'wavelength' λ of the perturbation is greater than

$(8\pi\rho_0)^{-\frac{1}{2}}$ the condensation will show a limited increase or decay while if $\lambda < (8\pi\rho_0)^{-\frac{1}{2}}$ there will be an oscillation with decreasing amplitude in an expanding universe.

As we have already remarked in the Newtonian discussion, the Lemaitre universe, in which there is a quasi-static epoch ($\dot{R} \ll 1$), may give an exponential increase of condensations with time. We first investigate some characteristics of the quasi-static phase. We recall the equations (3.8) and (3.9) (p. 24) putting $p = 0$, $k = 1$. $\Lambda \neq 0$.

$$\frac{8\pi\rho}{3} = \frac{1}{R^2} + \frac{\dot{R}^2}{R^2} - \frac{\Lambda}{3} \tag{12.38}$$

$$0 = \frac{1}{R^2} + \frac{\dot{R}^2}{R^2} + 2\frac{\ddot{R}}{R} - \Lambda . \tag{12.39}$$

For the quasi-static phase $\dot{R}$ is a minimum, so that $\ddot{R} = 0$, and $\dddot{R} > 0$, so that

$$\Lambda = 4\pi\rho_s = R_s^{-2}(1+\varepsilon) \tag{12.40}$$

where the subscript s refers to the values at the quasi-static phase and $\varepsilon = \dot{R}_s^2$. Using the matter conservation relation ρR^3 = constant, we get from equation (12.40)

$$\frac{4\pi\rho_0 R_0^3}{R_s^3} = \frac{1+\varepsilon}{R_s^2} , \tag{12.41}$$

the subscript zero referring to the present epoch. Introducing the density parameter σ_0 and the Hubble constant H_0, equation (12.41) reads

$$\sigma_0 = \frac{(1+\varepsilon)R_s}{3H_0^2 R_0^3} = \frac{1+\varepsilon}{3H_0^2 R_0^2 (1+z_s)} . \tag{12.42}$$

Again from equation (12.38)

$$2\sigma_0 = H_0^{-2} R_0^{-2} \left(1 - \frac{\Lambda}{3} R_0^2\right) + 1 \tag{12.43}$$

and substituting from equation (12.42) for $H_0^2R^2$, we get from equation (12.43)

$$\sigma_0^{-1} = (1+z_s)^3 - 3\,\frac{(1+z_s)}{(1+\varepsilon)} + 2\;. \qquad (12.44)$$

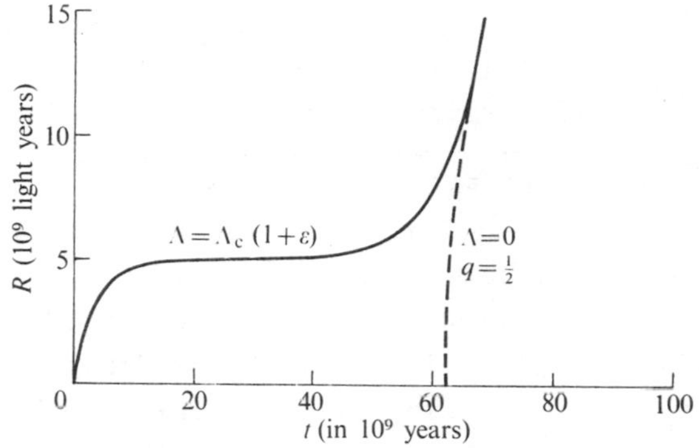

Fig. 12.1. $R(t)$ for $z_s = 1{\cdot}95$, $H_0 = 100$ Km.s^{-1}Mpc, and $\varepsilon = 2\text{x}10^{-5}$ (Kardshev, 1967).

This formula, given by Kardashev (1967) for $\varepsilon \approx 0$, requires modification when the pressure is not negligible. Considering a mixture of non-interacting pressureless matter and radiation, Brecher and Silk (1969) deduced the formula

$$\sigma_{om}^{-1} = (1+z_s)^3 - \frac{3(1+z_s)}{1+\varepsilon} + 2 + 2\alpha\left[1+(1+z_s)^4 - 2\,\frac{(1+z_s)^2}{(1+\varepsilon)}\right]$$

where α is the ratio of the radiation density and matter density at present and σ_{om} is the density parameter considering the matter density alone. The observational constraints on the possible values of σ_0 and α show that the radiation correction term is, in practice, quite negligible.

One may set an upper limit for z_s. Brecher and Silk considered the Oort value of $\rho_m \gtrsim 3{\cdot}1 \times 10^{-31}$ g cm^{-3} along with $H_0 = 75$ km s^{-1} Mpc^{-1} and obtained $z \lesssim 3{\cdot}2$. This was considered to be consistent with the fact that the observed red shifts of quasars showed a bunching near $z \sim 1{\cdot}95$ and at that time no quasar with a red shift z exceeding 3 was known, the idea being that the quasars were formed during the quasi-static epoch of a Lemaitre universe. However at present a value of $H_0 = 50$ km s^{-1} Mpc^{-1} seems more likely, this would scale down Oort's estimate to $\rho_m \gtrsim 1{\cdot}4 \times 10^{-31}$ g cm^{-3} and consequently $z_s \lesssim 1{\cdot}8$ and also quasars with red shifts greater than 3 have been observed. Thus the 'bunching' phenomenon in the quasar red-shifts cannot find an explanation in terms of the quasi-static phase of the Lemaitre universe.

Returning to the consideration of galaxy formation, Brecher and Silk (1969) calculated the duration of the quasi-static epoch τ_s as given by the approximate formula

$$\tau_s = H_0^{-1}\left[\frac{2+(1+z_s)^3-3(1+z_s)}{3(1+z_s)^3}\right]^{\frac{1}{2}} \ln \varepsilon^{-1} ,$$

and thus τ_s can be quite large if ε is sufficiently small. (τ_s is defined as the time during which R changes from $R_0\,(1+z_s)^{-1}(1-\delta)$ to $R_0(1+z_s)^{-1}(1+\delta)$ where $1 \gg \delta^2 \gg \varepsilon$.) As the quasi-static phase is still matter dominated, we may neglect the terms arising from pressure and hence any weak condensation would be multiplied by the factor $\exp[(4\pi\rho)^{\frac{1}{2}}\tau_s]$ during the phase. If the original perturbation is due to thermal fluctuation, then for the condensation to have the value unity at the end of the quasi-static phase, τ_s must have a quite high value ($\tau_s > 10^{10}$ years). Such a value of τ_s again requires a low value of ε or in other words, during the quasi-static phase the expansion rate must be extremely small. However the generation of a condensation leads to a conversion of matter into radiation with a consequent increase of the gravitational interaction as this depends on $\rho+3p$ rather than ρ. In fact, for the same energy-density, radiation is twice as effective gravitationally as matter. Brecher and Silk found that with the small value of $\dot{R}_s$, the increased gravitation leads to a reversal of the expansion and the universe as a whole starts collapsing. Thus the growth of a condensation, although possible, would lead to a collapse of the entire background universe.

It is thus clear that the Lemaitre universe does not offer an explanation of the formation of galaxies either if the original perturbation is due to thermal fluctuation. However Brecher and Silk found that for moderate values of ε ($\sim 10^{-6}$) it is only necessary to have a fluctuation of one part in 10^8 to explain the formation of galaxies, whereas the corresponding requirement is 10 per cent in case of Friedmann models with vanishing cosmological constant.

12.4. DENSITY PERTURBATIONS IN ANISOTROPIC MODELS

So far we have considered the background universe to have the Robertson-Walker line element. The problem of density perturbation in Bianchi type I universes have been studied by Johri (1972), Perko, Matzner, and Shepley (1972), and Bandyopadhyay (1977). The most thorough investigation is that by Perko *et al.* (1972) who found that the growth is complicated by the flow of energy between gravitational waves and the rudimentary condensations. They arrived at the result that the net effect is to modify the power-law exponent - from $t^{2/3}$ in an isotropic universe, it can range from t^0 to $t^{8/3}$ depending on direction and the particular type of model chosen. We shall, however, only give here the equation analogous to equation (12.31) with $\alpha = 0$, as given by Johri and Bandyopadhyay.

Proceeding exactly as before but remembering that the shear in the unperturbed state is now not zero, Bandyopadhyay found

$$\ddot{s} + \frac{4}{3t}\left(\frac{3t+2A}{3t+4A}\right)\dot{s} - \left(\frac{4}{3}A^2\rho^2+4\pi\rho\right)s = 0 ,$$

where the unperturbed density ρ and the shear σ^2 are given by

$$\rho = 4(3t^2+4At)^{-1}$$

$$3\sigma^2 = A^2\rho^2 \quad (A = \text{const.}).$$

The very same equation was obtained by Johri using a tetrad formalism. (See also Doroshkevich *et al.* 1971.)

12.4. DENSITY PERTURBATION IN BRANS-DICKE COSMOLOGY

Bandyopadhyay (1977) using the 'extended Raychaudhuri equation', deduced by Banerji (1974) for the Brans-Dicke theory, studied the linear density perturbation in the cosmological model given by equation (9.24) (p.172). He obtained two differential equations in which the perturbations of the density and the scalar field were coupled together. His final conclusion was that here too one has a power law for the growth of perturbations and thus was unable to explain the formation of galaxies as a result of thermal perturbations.

13
THE FORMATION OF GALAXIES

13.1. THE PRIMEVAL TURBULENCE THEORY

As the simple thermal fluctuation theory cannot give a satisfactory account of the formation of galaxies, it has been suggested that a turbulence in the early stage of the big-bang universe might have resulted in the formation of galaxies and cluster of galaxies (von Weizsacker 1951; Ozernoy and Chernin 1968, 1969; Chernin 1970, 1971, 1972; Ozernoy and Chibisov 1971; Sato, Matsuda, and Takeda 1970, 1971; Tomita, Nariai, Sato, Matsuda, and Takeda 1970; Silk and Ames 1972; Stecker and Puget 1972).

Consider the condition of the universe before the recombination of protons and electrons to form hydrogen. The cosmic fluid may then be considered to be a mixture of radiation with pressure $p_\gamma = \rho_\gamma/3$ and pressureless matter of density ρ_m. They are strongly coupled via the Thomson scattering but there is no interconversion. As $\rho_m R^3$ = const. and $\rho_\gamma R^4$ = const., we have $\rho_m \propto \rho_\gamma^{3/4}$ and the sound velocity c_s is given by

$$c_s^2 = \left(\frac{\partial p}{\partial \rho}\right)_s = \left(\frac{\partial p}{\partial \rho_\gamma}\right)\left(1 + \frac{\partial \rho_m}{\partial \rho_\gamma}\right)^{-1} = \frac{1}{3}\left(1 + \frac{3}{4}\,\frac{\rho_m}{\rho_\gamma}\right)^{-1} .$$

Thus in the radiation dominated phase, $c_s = 1/\sqrt{3}$ (remember the velocity of light is unity) and indeed so long as recombination does not occur c_s remains comparable with unity. After recombination, however, the matter is decoupled from radiation and the sound velocity drops to

$$c_s \sim (kT_m/m_p)^{\frac{1}{2}}$$

which is only a few km s^{-1}, this in our units, is a very small fraction of unity.

There are important questions as to how far the transition from the plasma state to the atomic-hydrogen state is abrupt and the speed of sound during the process of recombination. While Peebles (1968) finds that recombination is a very fast process, an energy input due to dissipation of turbulence may lead to the presence of a residual ionization for fairly long periods. Matsuda, Sato, and Takeda (1971) have given an empirical formula for the sound speed during recombination while Bonometto, Danese, and Lucchin (1974) have found that in the presence of a residual ionization, the velocity of sound is increased, the size of increase depending on the wavelength of sound. Even for small values of the residual ionization, the change of the turbulence from subsonic to supersonic may be significantly delayed and may delay or even prevent the formation of galaxies (Bonometto, Lucchin and Puget 1974; Bonometto *et al.* 1975).

In view of the high sound velocity in the pre-recombination era, the turbulence at these stages would be subsonic and the cosmic fluid is then practically incompressible. The relative velocity v_r between two points at a distance r apart follows Kolmogoroff's law $v_r \sim r^{1/3}$ and there is no large density fluctuation:

$$\frac{\delta\rho}{\rho} \sim \left(\frac{v}{C_s}\right)^2 ,$$

v being the turbulent velocity. When the sound velocity drops following recombination, the turbulence would become supersonic. The gas is now compressible and the Kolmogoroff law is modified to $v_r \sim r^n$, the exponent n depending on the compressibility. Large density fluctuations may now develop and provided they are energetically bound, they may lead to the formation of galaxies. The amount of turbulence that the separating condensations contain may account for the angular momenta of galaxies.

In order that the turbulent state may be in a stable equilibrium, the hydrodynamic time τ_h should be short compared to the cosmological time τ_{exp}

$$\tau_h < \tau_{exp} \tag{13.1}$$

or equivalently the turbulent velocity must be large compared to the cosmic expansion velocity,

$$v_{\mathrm{r}} > v_{\mathrm{exp}} \tag{13.2}$$

For the simplest case of the Einstein-de Sitter universe,

$$v_{\mathrm{exp}} = \frac{2}{3}\ rt^{-1} = \frac{2}{3}\, r_0(1+z)^{\frac{1}{2}}\ t_0^{-1}\ . \tag{13.3}$$

The inequality (13.2) and the equation (13.3) determine an upper bound for the scale of turbulence

$$r_0 < \frac{3}{2}\, v_{\mathrm{r}}\ t_0(1+z)^{-\frac{1}{2}}\ . \tag{13.4}$$

The Loytsianski theorem gives the conservation of angular momentum

$$J = (\rho_{\mathrm{r}}+\rho_{\mathrm{m}})r^4\ v_{\mathrm{r}}\ ,$$

so that in the radiation dominated phase one gets v_{r} = const. = V_{r} (say). However, in the matter dominated phase $\rho_{\mathrm{m}}R^3$ = const. and $\rho_{\mathrm{r}}+\rho_{\mathrm{m}} \approx \rho_{\mathrm{m}}$ so that $v_{\mathrm{r}} \sim (1+z)$. Hence we may write for the turbulence velocity in the matter dominated phase

$$v_{\mathrm{r}} = V_{\mathrm{r}}(1+z)(1+z_{\mathrm{e}})^{-1}\ , \tag{13.5}$$

where z_{e} is the value of z corresponding to the epoch $\rho_{\mathrm{m}} = \rho_{\mathrm{r}}$. The velocity V_{r} depends on the scale length. For the radiation era, one may take an empirical form

$$V_{\mathrm{r}} \sim r^{\alpha} \tag{13.6}$$

where $\alpha < 1$ and for the matter dominated phase one has the Kolmogoroff law

$$V_{\mathrm{r}} \sim r^{1/3}\ . \tag{13.7}$$

Combining (13.4) to (13.7), one finds that the upper bound to the turbulence-scale length increases in the radiation phase and decreases in the matter dominated phase, so that the bound has a maximum value at $z = z_{\mathrm{e}}$ (i.e. $\rho_{\mathrm{r}} = \rho_{\mathrm{m}}$). In particular, in the matter dominated phase the upper bound decreases as $(1+z)^{3/4}$.

However one has also a lower limit to the scale of turbulence, determined by the condition that the dissipation time due to viscosity must be greater than the cosmological time. While Ozernoy and Chernin (1968, 1969) as well as Sato *et al.* (1970, 1971) take the decay time τ_d, due to viscosity, of an eddy of size r as

$$\tau_d = r^2 \nu^{-1}$$

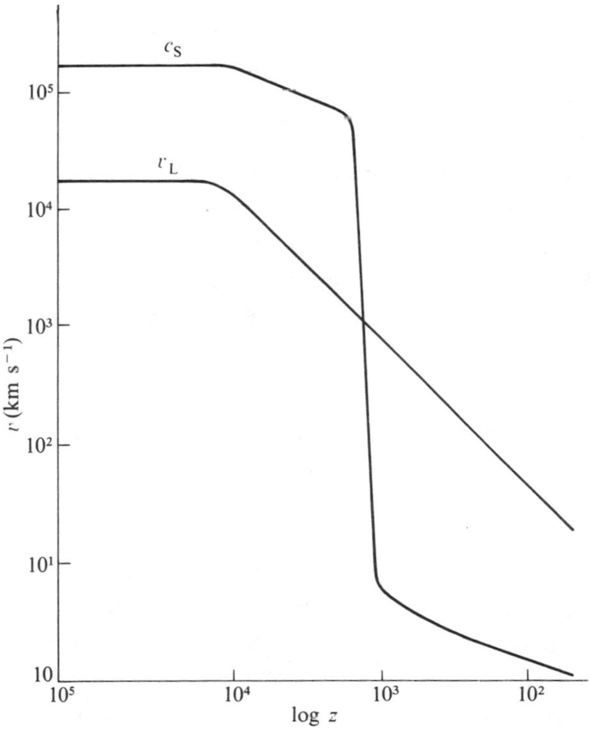

Fig.13.1. Sound velocity and turbulence velocity as function of z. (Silk, 1973b.)

where ν is the kinetic viscosity, and thus the limiting value of r is given by

$$\frac{r^2}{\nu} > \tau_{exp} .$$

Dallaporta and Lucchin (1972) have adopted a different relation for the lower bound following Landau and Lifshitz (1959).

This lower limit to the size of eddies rapidly increases as z decreases owing to the consequent increase of τ_{exp}. Thus the permitted range of turbulent eddies continually shrink in the matter dominated phase and there is a bottleneck at the recombination epoch. Dallaporta and Lucchin (1972) found that, even assuming the maximum possible value of the turbulent velocity (consistent with the subsonic condition) the bottleneck is rather narrow. This is essentially due to the fact that the radiation viscosity has a large value and is very effective in the decay of turbulence at almost all scales. However, they could apparently obtain a reasonably good fit with the data concerning the masses of the galaxies and the mean angular momentum per unit mass of spiral galaxies; the present matter density in this case is $\sim 0{\cdot}6 \times 10^{-29}$ g cm^{-3}. (The value of H_0 was assumed to be 75 km s^{-1} Mpc^{-1}.)

The situation becomes worse when one takes dissipation into account. Indeed Dallaporta and Lucchin (1973) and Jones (1973) found that with dissipation, galaxy formation is only possible for a very low density universe - the present density being not greater than 2×10^{-30} g cm^{-3}. This however may not seem very disturbing in view of the similar restriction on density demanded by the observational deuteron abundance (assuming the deuteron production to be cosmological) and also the presently favoured very low values of the deceleration parameter q_0. However, even with low densities, the origin of only spiral galaxies can perhaps be explained, no explanation seems to be forthcoming for the origin of elliptical galaxies of low specific angular momentum (see Anile, Danese, Zotti, and Motta 1976).

13.2. OTHER IDEAS ABOUT THE ORIGIN OF GALAXIES

Carlitz, Frautschi, and Nahm (1973) have suggested that in the hadron dominated big bang, one may have an explanation of galaxy formation without introducing considerations of turbulence if one takes the constant B in the hadron level density formula

$$N(m) \sim Am^{-B} \exp \beta m$$

to have the value -3 (Hagedorn 1965, 1970) instead of -2·5 (Nahm 1972). (See also Press and Schechter, 1974; Hamer and Frautschi, 1971.) This does not alter the limiting temperature and the non-relativistic condition of the particles at the big bang, but the matter remains in a non-equilibrium state throughout the hadron era. The matter has a grainy structure where each grain is a single massive resonance. The decay chains of these resonances may be so lengthy that the radiation era is considerably foreshortened with the following consequences: (i) Large density fluctuations occur in the hadron era are not damped out during the radiation era and thus one may have an explanation of galaxy formation. (ii) The model does not experience any difficulty in explaining the thermalization of the background radiation but the helium production as well as the deuteron abundance becomes a serious problem.

In the above model of galaxy formation, Carlitz *et al.* suggested that the mass $M(t)$ of a typically gravitationally bound lump increases as t, i.e. $M(t) \propto t$ and hence any grainy structure grew with time and ultimately reached the masses of galaxies, Peebles (1974b) has argued that the relation between M and t would be

$$M(t) \propto t^{4/7}$$

and hence the ratio of $M(t)$ to mass within the horizon varies as $t^{-3/7}$, therefore the large scale structure in the present universe could not have been due to grainy structure at too small a value of t.

Ryan (1972) suggested that the galaxies might have been formed by the accretion of matter about black holes of mass

10^7 M. Carr (1975) found that the number of such primordial black holes might indeed have been of the same order as the observed number of galaxies. Mészáros (1975) has claimed that if one assumes th presence of primeval black holes, their random statistical fluctuations may provide an explanation of galaxy formation.

One may pose the problem of galaxy formation in a different manner. The existence of galaxies with characteristic properties is a fact - can we calculate back the circumstances that would lead to the formation of such galaxies? Thus it has been suggested that the perturbations might be specified by a Gaussian random process in the form of a white power spectrum with a suitable cut off (Peebles 1967; Peebles and Dicke 1968). One can also study the properties that such galaxies must have in their formative period (Partridge and Peebles 1967; Tinsley 1972c,d, 1973b; Quirk and Tinsley 1973; Truran and Cameron 1971). Observations by Eggen, Lynden-Bell, and Sandage (1962) and Dixon (1966) have now apparently indicated that stars with high-space velocities and high-orbital eccentricities had been formed during the initial collapse of the protogalactic gas cloud and yet they contain some heavy elements. Hence there must have been a high initial rate of star formation or an initial stellar population of massive short-lived stars. In either case one would expect the early galaxies to have a high luminosity. Partridge (1974) and Davis and Wilkinson (1974) failed to detect any such high luminosity galaxy at large red-shifts (see, however, Larson 1974; Larson and Tinsley 1974; Meier 1976). For the roles of entropy fluctuation, baryon-antibaryon ratio fluctuation, and vacuum fluctuations of the gravitational field in the formation of galaxies, see Zel'dovich (1966), Harrison (1968), and Novello (1977b).

13.3. AN EXPLANATION OF GALACTIC MAGNETIC FIELDS

As has been pointed out by Harrison (1969, 1970a), turbulence in the radiation cum plasma fluid, bringing in vorticity, may generate a 'seed field'. The electron component of the plasma is strongly coupled with the radiation via Thomson scattering, while the positive ions may be supposed to expand independently as a pressureless gas. Then, if the electrons and positive ions have the same angular velocities at a certain stage, the electronic velocities would decrease as R^{-1}, whereas that of the positive ions would decrease as R^{-2} (cf. equations 12.22 and 12.23 on p.210). Thus there would be a differential rotation between the ions of opposite sign and consequently an electric current and a magnetic field.

One can calculate this 'seed field' assuming the turbulence to have occurred in the radiation dominated era. Let $\underline{V}_j = \underline{u}_j + \underline{v}_j$ be the mean velocity of ions ($j = b$ stands for positive ions, and $j = e$, $j = r$ for electrons and photons respectively). $\underline{u}$ is the expansion velocity and is irrotational and $\underline{v}_j$ is the turbulence velocity, assumed solenoidal. (In the radiation dominated phase, as we have already seen, the turbulence is subsonic and so density changes are negligible.) Following Harrison, we write down the equations (non-relativistic) for a fully ionized hydrogen plasma cum radiation

$$\rho_b R^{-1} \frac{D}{Dt} (R\underline{v}_b) = \frac{ne}{c} \left[\underline{E} + \underline{V}_b \times \underline{B} - \frac{\underline{J}}{\sigma}\right] \qquad (13.8)$$

$$\rho_e R^{-1} \frac{D}{Dt} (R\underline{v}_e) = -\frac{ne}{c} \left[\underline{E} + \underline{V}_e \times \underline{B} - \frac{\underline{J}}{\sigma}\right] + \frac{4\rho_r}{3t_{re}} (\underline{v}_r - \underline{v}_e) \qquad (13.9)$$

$$\frac{4}{3} \rho_r \frac{D}{Dt} \underline{v}_r = -\frac{c^2}{3} \underline{\nabla}\cdot\delta\rho_r - \frac{4}{3}\frac{\rho_r}{t_{re}} (\underline{v}_r - \underline{v}_e) + \frac{4}{3} \rho_r \nu_0 \nabla^2 \underline{v}_r \,, \qquad (13.10)$$

where n is the ion density of either sign, ρ the mass density and σ the electrical conductivity; ν_0 the radiation kinematic viscosity is given by $\nu_0 = c/5n\,\sigma_T$, σ_T being the Thomson cross-section and $t_{re} = 1/nc\,\sigma_T$ is the photon electron collision time. The derivative

$$\frac{D}{Dt} = \frac{d}{dt} + (\underline{v}_j\cdot\underline{\nabla}) = \frac{\partial}{\partial t} + (\underline{V}_j\cdot\underline{\nabla})$$

follows the total motion and d/dt follows the cosmological expansion.

Along with equations (13.8) – (13.10) we have the usual Maxwell equations

$$\underline{\nabla} \times \underline{E} = -\frac{\partial B}{\partial t}$$

$$\underline{\nabla} \times \underline{B} = 4\pi \underline{J}$$

Now taking the curl of equation (13.8) and using Maxwell's equations, we get

$$R^{-2}\frac{D}{Dt}[R^2(\underline{\omega}_b+\alpha\underline{B})] = [(\underline{\omega}_b+\alpha\underline{B}).\underline{\nabla}]\underline{v}_b + \alpha\eta\nabla^2\underline{B}$$

where $\alpha = e/m_p c$, m_p being the proton mass and $\eta = 1/4\pi\sigma$. $\underline{\omega}_b$ is the vorticity for positive ions $= \underline{\nabla} \times \underline{v}_b$. The first term on the r.h.s. vanishes in case of uniform rotation and the second term depends on the non-uniformity of the magnetic field. Neglecting these terms one obtains

$$\left(\underline{\omega}_b + \frac{e}{m_p c}\underline{B}\right) R^2 = \text{const.}$$

However for the radiation dominated phase one may take $\underline{\omega}_b \propto R^{-1}$. Thus even if $\underline{B}$ initially vanishes, it will finally approach a value

$$\underline{B} = -\frac{e}{m_p c}\underline{\omega}_b ,$$

and Harrison estimates this to have a value $\sim 10^{-14}$ G.

14
BARYON SYMMETRIC COSMOLOGY

14.1. THE KLEIN-ALFVÉN THEORY

It has often been held that as our basic physical theory is symmetric as far as particles and antiparticles are concerned, the universe should also exhibit this symmetry and thus the net baryon-number should vanish. It is true that in our neighbourhood there is an overwhelming preponderance of matter, the evidence against imagining that some of the galaxies are composed of antimatter is not compelling. Obviously spectroscopic data would not show any difference, however if one knows the direction of the magnetic field by some other means, the Zeeman effect would reveal the nature of the constituent of a distant body. Again, as we shall see, the Faraday rotation may provide a means of deciding the relative matter-antimatter ratio. A significant flux of radiation, if identifiable with annihilation radiation, may also provide evidence in favour of presence of antimatter. But before going to examine these observational evidences, let us have a look at the theories that have been developed assuming a net zero baryon number.

In the form proposed by Klein and developed by the Upsala school (Klein 1958; Alfvén 1965; Laurent and Soderholm 1969) the universe is supposed to have originated in the form of an extremely dilute cloud of equal concentrations of matter and antimatter, i.e. of protons, antiprotons, electrons, and positrons. This symmetric plasma (called the ambiplasma) starts as a homogeneous mixture. It is supposed that, due to gravitational instability, a certain part of this plasma becomes detached from this background and begins to contract and constitutes the so-called metagalaxy. As annihilation of matter and antimatter generates radiation, the metagalaxy becomes a mixture of plasma and radiation. It is then claimed that a spherically symmetric field of such a mixture contracts to a minimum and then bounces back if the mass of the metagalaxy is below a certain critical value ($M_{crit} \sim 10^{54}$ g.) One then has an expanding system showing a red shift progressively increasing with distance as in the observable universe.

It has been claimed that the theory, besides having an explicit symmetry between matter and antimatter, is also free from any singular state as occurs in conventional relativistic cosmology and explains, rather than assumes, the recession of the galaxies. A very ingenious mechanism for the separation of matter and antimatter has also been proposed. Consider the influence of a gravitational field on a symmetric ambiplasma. The barometric formula for the fall of density with height gives

$$n_{\mathrm{p}^-} = n_{\mathrm{p}^+} = A \exp(-\mu_{\mathrm{p}} z)$$

$$n_{\mathrm{e}^-} = n_{\mathrm{e}^+} = B \exp(-\mu_{\mathrm{e}} z)$$

where n_{p^-}, n_{p^+}, n_{e^-}, and n_{e^+} are the concentrations of antiprotons, protons, electrons, and positrons respectively and μ_{p} and μ_{e} are defined as

$$\mu_{\mathrm{p}} = gm_{\mathrm{p}}/kT$$

$$\mu_{\mathrm{e}} = gm_{\mathrm{e}}/kT$$

g indicating the intensity of the gravitational field in the negative direction of the z axis and m_{p} and m_{e} are the masses of the protons and electrons respectively. The temperature T is assumed to be constant throughout the plasma.

It is clear that because of the large proton-electron mass-ratio, the distribution of heavy particles will fall off much more rapidly with increasing z and hence the regions with large z will be richer in lighter particles, the heavier ones crowding in regions of low z. Suppose now a current flows in the ambiplasma in the direction of increasing z, i.e. antiparallel to g. The positive charges now move towards increasing z and as in the regions of large z there were only light particles, it is essentially the positrons which will move to the greater values of z. Similarly from the other side of the column, antiprotons will move towards the smallest z values. Thus, in the central region, there will be a preponderance of electrons and protons (i.e. matter) while the border regions on either end will be populated mainly by antiparticles. Hence a combination of gravitational field and electric current can apparently bring about a separation between matter and antimatter.

A simple calculation indicates the order of magnitude of masses that may be separated by the above mechanism. If a current J flows for a time t, the mass M separated would be $\sim m_{\mathrm{H}}\, Jt/e$, where m_{H} is the mass of the hydrogen atom and e the electronic charge. Now currents in the plasma would be associated with magnetic fields. If we consider an infinite cylinder of radius r, then the field B due to the current at the surface of the cylinder is $2J/cr$, so that

$$M = \frac{m_{\mathrm{H}} crtB}{2e} \approx 0 \cdot 5 \times 10^{-4}\, rtB.$$

With upper limits to t and B of $t < 10^{10}$ years and $B < 10^{-4}$ gauss, the above gives

$$M < 0 \cdot 5 \times 10^{9}\, r,$$

so that with $r \sim 10^{23}$ cm one gets $M < 10^{32}$ g, which is even less than the mass of the sun and so the matter-antimatter separation is too poor. As, however, the mass that separates out is proportional to the linear dimension r and the total mass increases as r^3, the process will be more efficient relatively for smaller values of r. Thus Alfvén points out that on a scale $r \sim 10^{17}$ cm one may have a high degree of separation. A co-existence of matter and antimatter leads, in general, to a rapid annihilation. However it is suggested that at the interface of matter and antimatter there may be a thin layer of very hot ambiplasma which prevents the annihilation (cf. the so called Leiden-frost effect in case of water over a very hot surface).

The general picture of the Klein-Alfvén theory is no doubt interesting but there remain questions which have not been satisfactorily answered. Thus the bounce of the contracting metagalaxy is thought to be brought about by the interaction with the annihilation radiation. However a simple conversion of matter into radiation would only accelerate collapse owing to the greater gravitational influence of radiation. The only way seems to be to introduce some sort of non-uniform conversion into radiation - how such a non-homogeneity originates does not seem clear. Again the Klein-Alfven picture is apparently irreconcilable with the observed thermal nature of the microwave background-radiation for, as we have already seen, the Hawking-Penrose theorem inescapably leads to geodetic incompleteness once one accepts the black-body nature of the microwave background.

The separation between matter and antimatter depends on the existence of directed gravitational and magnetic fields which would require a preferred direction in the metagalaxy. It is then a relevant question whether this would be consistent with the observed high degree of isotropy of the microwave background. Thus both the bounce and the separation of matter and antimatter depend on significant departures from homogeneity and isotropy - the existence of such departures is by no means observationally suggested, much less confirmed. Thus at present the Klein-Alfvén theory seems to be of only some historical interest. A much more viable symmetric cosmology has been proposed by Omnes and collaborators (Ómnes 1972; Aldrovandi, Caser, Ómnes, Puget, and Valladas, 1973; Aly, Caser, Ómnes, and Puget 1974; Aly 1974). We present a very short report on this theory in the next section.

It has been argued that if the antiproton life time is short (10^{-7} - 10^{-8} s) then the overwhelming presence of matter can be reconciled with an initial zero baryonic number (Cline *et al.* 1977; Demaret and Vandermeulen 1978), however, so far the experimental data on this point are inconclusive. (See however Ganguli *et al*, 1978.)

14.2. THE OMNES THEORY

In this theory the development of separate baryon and anti-baryon regions having masses which must be at least of the order of galactic masses takes place in three stages. In the first stage, which occurs at temperatures kT > 350 MeV (i.e. less than 10^{-5} s after the big bang) there is a large number of baryon-antibaryon pairs in equilibrium with the radiation field. If the interaction between a baryon and antibaryon be repulsive, then there may be a phase transition resulting in two phases of opposite non-zero baryon number. (However whether such a phase transition is consistent with the present-day theories and experimental data on baryon-antibaryon interaction remains controversial; see e.g. Bogdanova and Shapiro 1974.)

In the next stage, corresponding to 25 KeV $< kT <$ 350 MeV, ($t <$ 1600 s) an annihilation of the baryon-antibaryon pairs occur due to diffusion of baryons into antibaryon regions and vice versa. However as a result of annihilation, one has charged and neutral pions which finally decay to give a supply of neutrinos, high-energy photons, and electron-positron pairs. The photons give rise to a pressure which is somewhat higher on the concave side of the matter (or antimatter) droplets. This excess pressure opposes the diffusion. As due to the expansion of the universe, the temperature drops, a stage is reached when the annihilation pressure becomes comparable to the thermal pressure and the diffusion is practically stopped. (The annihilation pressure may generate turbulence resulting in galaxy formation, see Stecker and Puget 1972).

In the third stage, the temperature has fallen further so that annihilation pressure, which now dominates, causes an effective surface-tension at the boundaries of the matter (or antimatter) regions and consequently the drops of matter (or antimatter) have a tendency to coalesce thereby reducing the surface energy. It is claimed that in this way sufficiently large drops of matter are formed to account for the galaxies or cluster of galaxies. However if correlation of matter and antimatter is taken into account, the value obtained is too small for the baryon-photon ratio - the calculated value being 10^{-12} whereas the observed value is 10^{-8} (Aly *et al.* 1974). By introducing a small amount of turbulence, one can apparently raise the calculated value to the observed level.

If one considers neutron diffusion instead of proton diffusion, the annihilation is greatly enhanced and no sizeable separation of baryon and antibaryons would occur - the baryon-photon number ratio being less than 10^{-16} (Steigman 1974). There is also no primordial production of light nuclei in this symmetric cosmology.

14.3. THE OBSERVATIONAL SITUATION

The most notable fact on the observational side is that so far no antinucleons have been observed in cosmic rays. If at least some of the cosmic rays are of extragalactic origin the symmetrical cosmology would lead us to expect some antinucleons in these rays.

The annihilation again gives rise to high-energy photons ($h\nu \gtrsim 70$ MeV) and one may expect these to show up in the diffuse γ-ray background, thus providing indirect support in favour of symmetric theories. However there seems to be considerable uncertainty about the exact form of the observed γ-ray spectrum and also there would be an uncertain red-shift of the annihilation radiation. It is thus not possible to reach any definite conclusions from the observation of γ-rays and while there have been claims on the one hand that there is excellent agreement between observations and the predictions of symmetric cosmology (Stecker, Morgan, and Bredekamp 1971; Ómnes and Puget, 1974), Steigman (1973) on the other hand has concluded that the γ-ray data indicate a very low antimatter fraction unless the matter and antimatter are completely separated.

The last observational evidence that one may examine are data on Faraday rotation of polarized radio-waves from extragalactic sources (Berge and Seielstad 1967; Gardner, Morris, and Whiteoak 1969; Gardner and Whiteoak 1969; Seielstad and Weiler 1971; Mitton 1973). These have been variously interpreted (Fujimoto, Kawabata, and Sofue 1971; Reinhardt 1972; Rees and Reinhardt 1972; Mitton and Reinhardt 1972). While the rotation may originate in our galaxy, the intergalactic medium, or the source galaxy or perhaps partly in all three of them, the observed increase of averaged magnitude of rotation with the red shift z of the source suggests that the rotation takes place primarily in the intergalactic space.

Now the rate of Faraday rotation is $\alpha q.B.\lambda^2$ where

$q = n_{e^-} - n_{e^+}$ (the lepton residue of the ambiplasma), B is the component of the magnetic field parallel to the wave vector, and λ is the wavelength of the radiation considered. Thus for a pure-matter distribution the sign of the Faraday rotation would follow the sign of the magnetic field. Assuming that the intergalactic magnetic field has a random structure and that the scale length of these magnetic-field regions obey some probability law, it follows that the distribution of rotation for many sources of same z should be given by the normal distribution and the variance of the distribution should increase with z.

Thus, while for a pure-matter Friedmann universe the variance should increase steeply with z, Nelson (1973) found that the observational data seem to flatten out at and beyond $z = 1{\cdot}85$. This may be taken to be an evidence in favour of presence of antimatter in the region beyond $z \approx 1{\cdot}85$. However the magnitude of the effect and the possibility of alternative interpretations make this a rather weak piece of evidence.

15
SOME ASSORTED TOPICS

15.1. THE EXTRAGALACTIC RADIO SOURCES

In this chapter we propose to discuss some topics whose role in cosmology remains to some extent unclear. It may seem somewhat odd that we have not included the radio source observations in Chapter 4 on 'the analysis of observational data'. The reason, as will be presently clear, is that some doubts persist as to how far these radio observations are cosmologically significant.

Most of the galaxies are rather poor emitter of radio waves. However a small fraction, some 10^{-3} - 10^{-4} of the galaxies with visual magnitude $\approx -20{\cdot}5$ are powerful radio emitters, their radio luminosities exceeding those of normal galaxies by several orders of magnitude.

Owing to their large radio luminosity, these sources permit a study up to great distances (and hence to a remote past) but as their absolute radio luminosities vary considerably (in astronomical parlance 'the absence of a standard candle') it is not possible to determine their distances from the observed radio flux-densities.

Besides the radio galaxies, there are also the quasi-stellar objects. There is considerable controversy as to the origin of the high red-shift of these objects. However the majority of opinions seems to be in favour of regarding the red shifts as cosmological.

The survey of radio sources have been made in several frequency regions: 178, 408, 1400 - 1420, 2700, and 5000 MHz, and have been utilized to determine the number of sources N above some specified flux density S (defined as the power received per unit area per unit frequency range). As we shall presently see, for a uniform distribuiton of sources without evolution in a static Euclidean universe one has $N \sim S^{-3/2}$. In the analysis of radio-source surveys, one tries to fit the data with a relation of the form $N \sim S^{-\beta}$ and thus determine β. (More precisely one breaks up the log N - log S plot into straight-line portions and determine the slope $-\beta$ for different portions.)

For 178 MHz, one has the revised 3C, 4C, and North Polar surveys (Bennett 1962; Pilkington and Scott 1965; Gower, Scott and Wills 1967; Ryle and Neville 1962; Gower 1966). The 3CR data show that for high values of S ($\gtrsim 9 \times 10^{-26}$ W m^{-2} Hz^{-1}), the index $\beta = 1 \cdot 9 \pm 0 \cdot 1$. This corresponds to a deficit of sources in our neighbourhood. The North Polar survey of Ryle and Neville (1962) as well as the statistical analysis of the 4C survey records by Hewish (1961) indicate that the number counts tend to converge to a finite limit for low values of S. Ryle and Neville gives $\beta \approx 1 \cdot 3$ for $S = (0 \cdot 25 - 2) \times 10^{-26}$ W m^{-2} Hz^{-1}.

For the frequency 408 MHz, one has the 5C 2,3,4; B1, B 1S, and the Parkes surveys (Braccesi, Ceccarelli, Fanti, Gelato, Giovannini, Harris, Rosatelli, Sinigaglia, and Volders 1965; Braccesi, Ceccarelli, Fanti, and Giovannini, 1966; Grueff and Vigotti, 1968; Ryle and Neville 1962; Pooley and Kenderdine 1968; Pooley and Ryle 1968; Pooley 1969; Willson 1970; Bolton, Gardner, and MacKay 1964; Day, Shimmins, Ekers, and Cole 1966; Shimmins, Day, Ekers, and Cole 1966; Ekers 1969; Mills, Davies, and Robertson 1973). At this frequency there is also the same deficit of sources at high flux-densities ($\beta \sim 1 \cdot 9$ for $S > 1 \times 10^{-26}$ W m^{-2} Hz^{-1}) while there is a marked convergence for $S < 0 \cdot 1 \times 10^{-26}$ W m^{-2} Hz^{-1}, β being about $0 \cdot 8$.

For the 1400 - 1420 MHz region (Galton and Kennedy 1968; Bridle and Davis 1972; Fitch, Dixon, and Kraus 1969; Harris and Kraus 1970, 1971; Price and Milne 1965; Bridle and Davis 1972; Fomalont, Bridle, and Davis 1974; Maslowski 1973; Katgert, Katgert-Merkelijn, Lepoole, and Van der Laar 1973) one has again an initial value of $\beta \approx 1 \cdot 9$ and a convergence at low flux densities.

For 2700 MHz, Shimmins *et al.* (1966) found $\beta = 1 \cdot 74 \pm 0 \cdot 15$ for $S > 1 \cdot 5 \times 10^{-26}$ W m^{-2} Hz^{-1} and $\beta \approx 1 \cdot 38$ for $S \leq 0 \cdot 4 \times 10^{-26}$ W m^{-2} Hz^{-1}. At 5000 MHz (Davis 1971; Pauliny-Toth and Kellerman 1972a,b 1974; Kellermann, Pauliny-Toth, and Davis 1968) also, the source densities show a significant drop in our neighbourhood (i.e. for high flux densities).

While the radio sources are generally distributed isotropically (Holden 1966; Wagoner 1967; Golden 1974; Pearson

1974) with little clustering (Hinder and Branson 1969; Hughes and Longair 1967; Webster 1976), it has been claimed that the GA and GB surveys at 1400 MHz show distinct anisotropy (Maslowski 1973; Maslowski, Machalski, and Zieba 1973; Machalski, Zieba and Maslowski 1974). This anisotropy was thought to have some link with a difference in the spectral index in different angular regions of the sky observed by Shimmins and Bolton (1974). However more recently the existence of anisotropy in the distribution has been questioned by Blake (1976).

Thus the surveys at different frequencies apparently more or less agree in indicating a non-uniformity in the form of what has sometimes been described as a 'hole' in our neighbourhood, where there is a marked paucity of sources. However it can hardly be considered to be a local fluctuation effect as the deficit in the number of sources extend to $z \approx 0{\cdot}3$. We have in the intermediate region after this 'hole' a fairly rich number of sources and again beyond about $z \sim 2{\cdot}5$, the number of sources decrease rapidly.

Before going to the theoretical analysis of this apparent non-uniformity, we look to other tests in this regard. As the observations extend up to a limiting flux density, many weak sources will be cut off at large distances, i.e. large z-values, while they would be detected at small values of z. Thus the N-z studies are of little value in deciding the question of uniformity of the number density. Again due to the large spread in the absolute luminosity of these sources, the N-m studies also do not provide any test of uniformity in the distribution.

These difficulties are obviated in the luminosity-volume test due to Schmidt (1968) and Rowan-Robinson (1968). Suppose that the survey of a sample of objects is complete in the sense that all objects having a flux density above a certain limiting value at a particular frequency have been detected; we may now compute for each source the limiting red shift z_{m} at which it would be just included in our sample. We may calculate the volume V_{m} corresponding to the red shift z_{m}. Again there is the volume V giving the volume up to the actual red shift z of the source. As $0 \leq z \leq z_{\mathrm{m}}$, we shall have V/V_{m}

lying between 0 and 1. Now if the distribution of the sources be uniform V/V_m would have a uniform distribution in the range and for a complete sample the average value $\langle V/V_m \rangle$ would be 0·5. Considering the 3CR survey at 178 MHz Rowan-Robinson (1971) found that taking account of all sources $\langle V/V_m \rangle$ = 0·631 which indicates significant departure from uniformity. A break up of the sources into two groups showed that the departure from uniformity could be attributed entirely to 'stronger radiogalaxies'.

To compare the observational results with theoretical expectations from the isotropic models, we must remember that the absolute luminosities have a wide dispersion. One thus introduces a function ρ such that $\rho[P(\nu)]dP(\nu)$ is the number of sources per unit proper volume with power between $P(\nu)$ and $P(\nu) + dP(\nu)$ in the frequency range ν to $\nu+d\nu$. Owing to the expansion of the universe, the proper volume containing a particular number of sources varies as R^3, hence ρ is not a constant even if there be a conservation of the sources. In the latter case we have

$$\rho_e = \rho_0 \frac{R_0^3}{R_e^3} = \rho_0(1+z)^3 ,$$

where the subscripts 0 and e correspond to the epochs of observation and emission respectively.

The sources are assumed to have a power spectrum

$$P(\nu) \propto \nu^x \tag{15.1}$$

where x is called the spectral index. The spectrum would have a cut-off on the high frequency side. If ν_0 be the observation frequency, then for a source showing the red shift z, the emitted frequency $\nu_e = \nu_0(1+z)$ and

$$P(\nu_e) = P(\nu_0)(1+z)^x . \tag{15.2}$$

The flux density S as observed for this source is

$$S = P(\nu_0)(1+z)^{1+x} D^{-2} , \tag{15.3}$$

where D is the luminosity distance of the source. An additional factor $(1+z)$ has appeared in equation (15.3) as the emitted frequency range $d\nu_e = d\nu_0(1+z)$, $d\nu_0$ being the observed frequency range. The number $N(> S_0)$ of sources giving a flux density $\geq S_0$ at the observer is

$$N(> S_0) = 4\pi R_0^3 \int_0^r \left(1 + \frac{Kr^2}{4}\right)^{-3} r^2 \int_{P_0}^{\infty} \rho_0[P(\nu_0)]dP(\nu_0)dr \quad (15.4)$$

where in view of equation (15.3),

$$P_0 = S_0 D^{+2}(1+z)^{-1-x} \quad .$$

In equation (15.4) the upper limit of r is that for which the cut-off frequency is red-shifted to the frequency of observation. Beyond this a source will not be included in the observation. For the case of static Euclidean space,

$$k = z = 0$$

$$D = r\ R_0$$

$$P_0 = S_0\ r^2\ R_0^2 \ ,$$

and we get from equation (15.4)

$$N(> S_0) = \frac{4\pi}{3} S_0^{-3/2} \int_0^{\infty} P_0^{3/2}\ \rho_0(P_0)dP_0 \ ,$$

and one has the $\beta = 3/2$ law independent of the nature of the distribution function ρ_0.

For expanding relativistic models, the determination of the N-S relation from equation (15.4) would require a knowledge of the function ρ_0. A simplifying assumption is to consider sources to be of the same power, i.e. to assume that ρ_0 is of the δ-function form. One can express D and r in terms of z and the deceleration parameter q and then evaluate the integral of equation (15.4) in terms of z to get the N-S relation. The result of such calculations is that, irrespective of the type of model, β for high flux region (i.e. $z \to 0$) should be less than 1·5 (Ringenberg and McVittie 1969) whereas we have just seen that observations lead to $\beta \sim 1{\cdot}8$ for this region.

This discrepancy can of course be taken to mean that the universe is not really homogeneous and isotropic. However the evidence of microwave background cannot be brushed aside and considerations of simplicity, as well, make an alternative explanation more attractive. One thus assumes an evolution of the sources. The evolution may be of different types - it may be a non-conservation of the source number (i.e. a disappearance and appearance of sources) or it may be a change in the luminosity of the sources progressively with time or even a change in the spectral index with time. Indeed one may also conceive different combinations of these three. Unfortunately at the present state of our knowledge, it does not seem possible to specify uniquely the type of evolution involved and different authors have shown that the observational data could be well reproduced by different forms of evolution (Fanaroff and Longair 1973; van Hoerner 1973). Thus the N-S studies, while apparently ruling out the steady-state cosmologies as they do not admit any evolution, nevertheless do not bring any support for relativistic cosmology, much less do they select out any of the numerous possible models allowed by general relativity.

There has also been some studies on the angular sizes of the radio sources. However, here also difficulty arises due to the very wide range over which the sizes of the radio sources vary - the variation may be even by a factor 10^5. Again many sources have a 'hot spot' of typical size 5 kpc (Readhead and Hewish 1976). It has often been held that the evolution effects are more predominant in the sizes of the radio sources than possible geometrical cosmological effects and one may not therefore expect to find the minimum in the θ-z curve which is characteristic of isotropic relativistic cosmological models (Longair 1971). Nevertheless in a recent analysis of the angular diameter vs flux-density plot, Swarup (1975) and Kapahi (1975) have claimed to have found a minimum after correcting the sizes by a particular evolution formula. The evolution assumed is however rather arbitrary and has been questioned by Narlikar and Chitre (1977). (See also Dodd, Morgan, Nandy, Reddish, and Seddon 1975; Ellis, Fong, and Phillips 1976.)

5.2. MACH'S PRINCIPLE

We had occasion to refer to Mach's principle earlier on. We now present a short discussion of the various views on this principle. A very readable account is due to Reinhardt (1973).

Mach's principle may be said to have originated from the results of Foucoult pendulum and gyroscopic experiments. If the relative motion between two frames of reference are not uniform, then Newton's laws of motion in their simple form cannot hold in both the frames. If they do hold in one frame, called the inertial frame, then in the other, fictitious forces appear. If the motion is one of uniform rotation these forces are known as Corioli's and centrifugal forces and are proportional to the angular velocity ω and ω^2 respectively. The Foucoult experiments demonstrated the existence of these forces for an earth bound frame showing that it is not an inertial frame. The experiments went further and gave a value for the rotational velocity of the earth relative to the inertial frames.

It is also possible to determine a value for the rotational velocity of the earth from the observation of distant stars assuming that their apparent diurnal motion is due to the rotation of the earth. These two velocities, i.e. that determined from the Foucoult experiments and the other determined from astronomical observations, were found to be in agreement within the range of observational uncertainty. It can thus be concluded that the inertial frame is coincident, at least approximately, with the rest frame of distant heavenly bodies. Later observations showed that relative rotation, if any, between the inertial and the astronomical frames must be less than about 10^{-18} s^{-1} (Clemence 1957; Schiff 1964).

Again observations of the isotropy of the microwave and X-ray backgrounds indicate that the cosmological vorticity must be less than 10^{-25} s^{-1} (Collins and Hawking 1973b;

Schwartz 1970).

Is this remarkable coincidence between the two frames just accidental? Mach's answer was no. He proposed that the inertial frame is, in fact, determined by the distribution of masses in the universe. If this is so, one may rephrase Newton's laws as stating that the inertia of a body is nothing but its resistence to be accelerated relative to the overall matter distribution in the universe. If we imagine now an isolated particle in an otherwise empty universe, then no inertial frame is defined and thus inertia has no meaning. Again if we change the distribution or magnitude of the masses in the universe, the inertia of bodies should undergo a change. In short, inertia of a body is not its intrinsic property but is a result of interaction of other bodies of the universe with itself.

This idea leads one to expect a 'dragging' of the inertial frame by massive bodies in accelerated motion. Some of these expectations are apparently, at least to some extent, fulfilled in the general theory of relativity. Thus the calculations by Thirring and Lense (1918) brought out a dragging of the inertial frame inside a thin rotating spherical shell. However they used a weak field approximation. More recently Lansberg (1969) has shown up to first order in the angular velocity ω, the dragging effect of a rotating spherical shell (situated in the Einstein static universe) on the inertial frame at the centre, decreases with the distance and increases with the thickness of the shell, and the dragging is complete when, in the limit, the shell covers the whole universe.

Also, Einstein (1950) came to the conclusion that there is an increase of inertia of a body if matter is piled up in its neighbourhood but his calculation appears to be erroneous (Brans 1962). The result that inside a non-rotating spherical shell the metric is Euclidean seems to be in direct conflict with Machian ideas.

It has been argued that Mach's ideas involve a denial of the existence of any absolute space - space time, so to say, is generated by the matter distribution. Thus one may expect that there would be no everywhere regular solution of the field equations for a hypothetical absolutely empty universe. But the general relativity equations not only admit the Minkowski space but also other solutions (e.g. Taub 1951) and in case there is a non-vanishing cosmological constant, one has the de Sitter metric. Again one might demand from Mach's principle that there should be no rotating cosmological solution but a variety of such solutions exist as we have already seen.

It has sometimes been suggested that Mach's principle may mean the imposition of some boundary conditions which would exclude these un-Machian solutions. This is conceivable for, in general relativity the $g_{\mu\nu}$ depend not only on the $T_{\mu\nu}$ s but also on the boundary (or symmetry) conditions. Wheeler (1964a,b) put forward the idea that Mach's principle required the universe to be closed and singularity free. However singularity-free realistic cosmological solutions do not exist and recent observations seem to favour open models. A background dependent approach towards the Mach principle has been considered by Gürsey (1963) and recently by Goldoni (1976).

Some recent attempts to select solutions consistent with Mach's principle have been based on the possible replacement of the Einstein field equations by integral equations as the electromagnetic wave equations may be transformed to the Kirchoff integral form. Following the earlier investigations of Al'tshuler (1967, 1969) and Lynden-Bell (1967), Sciama, Waylen, and Gilman (1969) showed that one could express $g^{\mu\nu}$ at any point as a volume integral over the sources $T_{\mu\nu}$ in the past light-cone of the field point plus a surface integral. The volume integral is of the form

$$\int G^{\mu\nu}_{\alpha'\beta'}\, T^{\alpha'\beta'}\, dx'$$

where the coefficients $G^{\mu\nu}_{\alpha'\beta'}$ (covariant with respect to the primed indices at the source point and contravariant with respect to the unprimed indices at the field point) themselves involve the $g^{\mu\nu}$'s and thus the basic non-linearity of the Einstein equations is taken care of. The surface integral can again be split up into a part due to sources outside the volume and a source-free part. Gilman (1970) interpreted the Mach principle as demanding the vanishing of the source-free part. Obviously the empty space case would not give any acceptable solution (one then gets $g^{\mu\nu} = 0$). However the Gilman criterion made many well known solutions anti-Machian.

Raine (1975) has found a relation between the metric tensor and the Riemann tensor and also another between the Riemann tensor and the energy-momentum tensor. As in Machian conditions he insists that the first relation must be unique up to a gauge transformation while the second one must be linear in the energy-momentum tensor. He thus finds that all empty space-times are non-Machian but the Robertson-Walker solution is Machian. A further result to which we have already referred is that the perfect fluid homogeneous universes with rotation are non-Machian; the homogeneous anisotropic universes belong to the same category. For inhomogeneous dust-universes, those with Robertson-Walker type singularities are Machian while others having Heckmann-Schucking type or Kasner type singularities are non-Machian.

It has often been claimed that the steady state cosmology or the Brans-Dicke field equations are more consistent with Mach's ideas. Indeed Gödel-type rotating cosmological solutions are apparently not admitted in these cases (Banerji 1974) but empty-space solutions are representing plane waves as also other rotating solutions exist (Bandyopadhyaya 1978a) similar to the Van Stockum (1937) solution in general relativity. However there are singularities and closed time-like lines in both Bandyopadhyaya and Van Stockum solutions. The Maitra (1966) solution in general relativity is free from these features and to this no analogue in Brans-Dicke theory exists (Bandyopadhyaya 1978b).

All the discussions so far have been at a purely theoretical level. Following an idea of Cocconi and Salpeter (1958), Hughes, Robinson, and Bettrow-Lopez (1960) attempted to detect a Machian effect. The idea is that if the distribution of matter about a body be anisotropic, the inertia of the body will exhibit a corresponding anisotropy. This of course involves the assumption that the Machian contribution to the inertia is not a scalar. Assuming this contribution to the inertia due to a mass M at a distance r to be of the form $Mr^{-\nu}$ (except for the angle dependence) Cocconi and

Salpeter (1958) concluded that owing to the anisotropy of the mass distribution in our galaxy, the inertia of a body should show a relative anisotropy of magnitude

$$\frac{\Delta m}{m} \approx \frac{M}{r^{\nu}} \frac{3-\nu}{4\pi\rho R^{3-\nu}} \quad ,$$

where M and r are related to the mass and dimension of the galaxy, ρ the averaged out density of matter in the universe and R the distance up to which Machian influence persists (for relativistic models this is limited by the horizon, i.e. $R = cH_0^{-1}$). Taking $\rho \approx 10^{-29}$ g cm^{-3}, $\nu = 1$, Hughes *et al.* estimated $\Delta m/m \sim 10^{-5}$.

It was argued that if inertia be direction dependent, the energies of the atomic and nuclear states would be affected. Thus the energy of an electronic state in the presence of a magnetic field would depend not only on n, j, and m, but also on the orientation of the magnetic field relative to the direction of the galactic axis. Thus the Zeeman splitting should depend on this orientation. Again for the ^{7}Li nucleus with spin 3/2, there are four energy levels in a magnetic field. If the inertia be isotropic, these levels would be equally spaced and consequently there would be a single nuclear resonance line. However with an anisotropic mass, the levels are no longer equi-spaced and there should be three resonance lines. From an apparent absence of such a triplet Hughes *et al.* concluded $\Delta m/m \leq 10^{-20}$, which is greatly below the expected Machian value.

Two points however deserve mention. The Machian contribution to inertia may well be scalar and in that case there would be no mass anisotropy. Secondly, as Dicke (1962b) has argued, the anisotropy in inertia should not be limited to particles alone but all field energies would be similarly affected. The inertia tensor would be of the form

$$m_{ij} = mf_{ij}$$

where the tensor f_{ij} is the same for all particles and fields. The variational principle

$$\delta \int m.\sqrt{(f_{ij}u^i u^j)}.\ ds = 0 ,$$

yields the geodesic equation with f_{ij} as the metric tensor. The anisotropy thus appears as a new metric.

The tensor f_{ij} would enter into the Lagrangian of fields as well and it then follows that there would be no observable effect on the motion and the energy states. Put in a different way the anisotropy being common to all particles and fields becomes absorbed in the definition of the metric tensor and one recovers the usual equations of motion.

15.3. OLBERS' PARADOX

Olbers' paradox is concerned with the background-radiation density, more commonly called the illumination of the night sky. In the universe we see the stars and the galaxies continuously pouring out radiation. This would go up to build a background-radiation density throughout the whole universe - of course for this it is necessary that the number of sources (or rather the amount of radiation poured out) must have a finite density considering the whole universe. Assuming that the average rate of emission of radiation per unit volume is U, then the growth of average radiation density ρ can be expressed by

$$U\ V = \frac{\mathrm{d}}{\mathrm{d}t}\,(\rho V) + \alpha\rho \qquad (15.5)$$

or,

$$\frac{\mathrm{d}\rho}{\mathrm{d}t} = U - \frac{\alpha\rho}{V} .$$

Here the last term represents the absorption of radiation; α thus involves both the absorption in the surfaces of stars and galaxies (including also dark objects) as well as in any interstellar and intergalactic gas or other matter that may be present.

Equation (15.5) will have an asymptotic equilibrium solution

$$\rho = \frac{UV}{\alpha}, \tag{15.6}$$

when the absorption exactly balances the emission. In particular if one considers the emission and absorption as surface phenomena at the surfaces of say the stars, the radiation density at the surfaces should be equal to the radiation density throughout the space outside, or, as the Olbers's paradox is commonly stated, the whole night sky should look uniformly bright and we should not be able to discern any distinct object like the stars.

In the above argument the assumptions that are involved are

(i) Sufficient time has elapsed to enable the radiation density to attain the value (15.6). To be more precise the radiation must be pouring out for a time $T \gg V/\alpha$.

(ii) The emission and absorption have remained constant during the time T. This assumption can be relaxed to a considerable extent but cannot be completely removed.

(iii) In the energy balance of radiation the only effective processes are emission and absorption.

In the usual solution of Olbers' paradox, proposed since the advent of expanding cosmological models of general relativity, it was pointed out that assumptions (i) and (iii) are both wrong because the universe had a beginning at a finite time in the past and this age of the universe might well be less than V/α. Again the red shift means a continuous degradation of the radiation energy and this invalidates the assumption (iii). However as Harrison (1964, 1965) has pointed out it is the invalidity of assumption (ii) rather than that of (i) and (iii) which is responsible for the darkness of the night sky. His argument rests on the fact that any radiation of energy means a loss of mass and hence a star or a galaxy with a finite mass can radiate only for a limited time. Substituting values, he shows that while the time required for the sky to attain the surface brightness of stars is 10^{23} years, an average star would spend all its mass in $\sim 10^{13}$ years if it radiates at the average rate of a main sequence star. He makes this argument, based essentially on the conservation of energy, look simpler and stronger by remarking that if the entire mass of the universe ($\sim 10^{-30}$ g cm^{-3}) be converted to radiation, the radiation temperature would only be about 30 K.

One may base the reasoning on the two laws of thermodynamics as well. If the universe had existed for a sufficiently long time it should have attained thermal equilibrium which, according to the second law of thermodynamics, corresponds to a uniform temperature throughout. Thus the radiation density should also be uniform. The answer again is that the time for attaining thermodynamic equilibrium is apparently much greater than the time that has so far lapsed and if indeed equilibrium were attained the radiation would have a very low density in view of the conservation of energy and the relatively poor amount of total energy available.

15.4. CONCLUDING REMARKS

As we come to the end of the book, it is natural to look back and ponder as to how much of systematization of observational data, our theories and speculations will stand the test of time and how much is likely to be swept away in the not too distant future. For long we have believed the red shift - apparent magnitude correlation to be basically correct and taken it to be the key to our understanding of the universe. But today that belief, to some extent at least, is shaken; evidences in favour of so-called anomalous red shifts seem increasingly difficult to be brushed aside and when one considers the quasar red shifts, there is almost complete confusion.

Is the idea of an expanding universe then just a fantasy and the big-bang a myth as Alfven (1977) would like to make us believe? Has the red shift an altogether different origin, or is the red shift a combined effect of recession and some other influence? Is the linear relation between the red shift and distance ($z \propto r$) (which we have believed so long) to give way to a quadratic relation ($z \propto r^2$) (Nicoll and Segal 1975) and are we to change over to some altogether new framework like the chronometric cosmology of Segal (1975, 1976)?

The evidence regarding the thermal nature of the microwave background is indeed to some extent weak, for the intensity in crucial wavelength region has not been measured. But the isotropy of the radiation, at our locale, though not exact, is to say the least remarkable. It is not easy to see how one can reconcile this characteristic of the radiation with anything other than a dense hot universe in the past.

If the idea of an expanding universe be not wrong, then what is the value of the deceleration parameter q_0? With estimates of q_0 ranging all the way from negative to positive values, high enough to make the universe closed, a large group of cosmologists seem to be converging to a philosophy of despair that we shall never be able to tell the value of q_0 with any reasonable confidence. If the history of science or indeed that of humanity be any guide, such pessimism seems to be ill founded.

Should however we be able to determine q_0 accurately, will it be consistent with Einstein's original equations? Or shall we be forced to resurrect the cosmological term? Will the value of q_0 be consistent with the time scale arrived from nucleocosmochronology and the study of globular clusters? Will it demand a renewed search for indirectly observed matter and eliminate the idea of deuteron production in the early universe?

So we end with a note of uncertainty but not of despair. At this moment the situation is to some extent confusing and it seems best to keep an open mind. There have been surprises in the past and let us hope there will be thrills in the future of cosmology as well.

APPENDIX

CONSTANTS OF STRUCTURE C^a_{bc}, THE KILLING VECTORS η^i_a , AND THE VECTORS ξ^i_a FOR THE NINE BIANCHI TYPES (Taub, 1951.)
In the following 1, 2, and 3 refer to x,y,z respectively.

Type I $C^a_{bc} = 0$ for all values of a,b and c.

$$\eta^i_a = \xi^i_a = \begin{bmatrix} 1 & 0 & 0 \\ 0 & 1 & 0 \\ 0 & 0 & 1 \end{bmatrix}$$

Type II $C^1_{23} = - C^1_{32} = 1$, all other $C^a_{bc} = 0$

$$\eta^i_a = \begin{bmatrix} 0 & 0 & 1 \\ 1 & 0 & z \\ 0 & 1 & 0 \end{bmatrix} \qquad \xi^i_a = \begin{bmatrix} 0 & 0 & 1 \\ 1 & x & 0 \\ 0 & 1 & 0 \end{bmatrix}$$

Type III $C^1_{13} = - C^1_{31} = 1$, all others zero.

$$\eta^i_a = \begin{bmatrix} 0 & 0 & 1 \\ 1 & 0 & y \\ 0 & 1 & 0 \end{bmatrix} \qquad \xi^i_a = \begin{bmatrix} 0 & 0 & 1 \\ e^x & 0 & 0 \\ 0 & 1 & 0 \end{bmatrix}$$

Type IV $C^1_{13} = - C^1_{31} = C^1_{23} = - C^1_{32} = C^2_{23} = - C^2_{32} = 1$; others zero.

$$\eta^i_a = \begin{bmatrix} 0 & 0 & 1 \\ 0 & 1 & y+z \\ 0 & 1 & z \end{bmatrix} \qquad \xi^i_a = \begin{bmatrix} 0 & 0 & 1 \\ e^x & xe^x & 0 \\ 0 & e^x & 0 \end{bmatrix}$$

Type V $C^1_{13} = - C^1_{31} = C^2_{23} = - C^2_{32} = 1$, all other $C^a_{bc} = 0$.

$$\eta^i_a = \begin{bmatrix} 0 & 0 & 1 \\ 1 & 0 & y \\ 0 & 1 & z \end{bmatrix} \qquad \xi^i_a = \begin{bmatrix} 0 & 0 & 1 \\ e^x & 0 & 0 \\ 0 & e^x & 0 \end{bmatrix}$$

Type VI $C^1_{13} = -C^1_{31} = 1$; $C^2_{23} = -C^2_{32} = h$ $(h \neq 0, 1)$, other $C^a_{bc} = 0$.

$$\eta^i_a = \begin{bmatrix} 0 & 0 & 1 \\ 1 & 0 & y \\ 0 & 1 & hz \end{bmatrix} \qquad \xi^i_a = \begin{bmatrix} 0 & 0 & 1 \\ e^x & 0 & 0 \\ 0 & e^{hx} & 0 \end{bmatrix}$$

Type VII $C^1_{23} = -C^1_{32} = -1$, $C^2_{23} = -C^2_{32} = h$ $(h^2 > 4)$, other $C^a_{bc} = 0$

$$\eta^i_a = \begin{bmatrix} 0 & 0 & 1 \\ 1 & 0 & -z \\ 0 & 1 & y+hz \end{bmatrix} \qquad \xi^i_a = \begin{bmatrix} 0 & 0 & 1 \\ p & 0 & 0 \\ 0 & q & 0 \end{bmatrix}$$

where

$$p = (h^2-4)^{-\frac{1}{2}} \, [(\alpha+1)e^{\alpha x} - (\beta+1)e^{\beta x}]$$

$$q = -(h^2-4)^{-\frac{1}{2}} \, [(\beta+1)e^{\alpha x} - (\alpha+1)e^{\beta x}]$$

α, β being roots of the quadratic $x^2 - hx+1 = 0$.

Type VIII $C^1_{12} = -C^1_{21} = 1$; $C^2_{13} = -C^2_{31} = 2$; $C^3_{23} = -C^3_{32} = 1$; other $C^a_{bc} = 0$.

$$\eta^i_a = \begin{bmatrix} e^{-z} & 0 & 0 \\ -y^2 e^{-z} & 0 & e^z \\ 2ye^{-z} & 1 & 0 \end{bmatrix} \qquad \xi^i_a = \begin{bmatrix} 0 & -x & x^2 \\ 0 & y & 1-2xy \\ 0 & 1 & -2x \end{bmatrix}$$

Type IX $C^1_{23} = -C^1_{32} = C^2_{31} = -C^2_{13} = C^3_{12} = -C^3_{21} = 1$, others zero.

$$\eta^i_a = \begin{bmatrix} 0 & \cos y & -\sin y \\ 1 & -\cot x \sin y & -\cot x \cos y \\ 0 & \sin y / \sin x & \cos y / \sin x \end{bmatrix}$$

$$\xi^i_a = \begin{bmatrix} -\sin z & \cos z & 0 \\ \cos z/\sin x & \sin z/\sin x & 0 \\ -\cot x \cos z & -\sin z \cot x & 1 \end{bmatrix}$$

REFERENCES

ADAMS, P.J. and CANUTO, V. (1975). *Phys.Rev.* D12, 3793.

AICHELBURG, P.C. (1973). *Phys.Rev.* D8, 377.

ALDROVANDI, R., CASER, S., and OMNES, R. (1973). *Nature, Lond.* 241, 340.

——— ——— ——— and PUGET, J.L. (1973). *Astron.Astrophys.* 28, 253.

ALEXANIAN, M. (1975). *Phys.Rev.* D11, 722.

——— and MEJIA-LIRA, F. (1975). *Phys.Rev.* D11, 715.

ALFVEN, H. (1965). *Rev.mod.Phys.* 37, 652.

——— (1977). *Cosmology, history and theology*, p.12. Plenum Press, New York.

ALPHER, R.A., BETHE, H., and GAMOW, G. (1948). *Phys.Rev.* 73, 803.

——— and HERMAN, R.C. (1949). *Phys.Rev.* 75, 1089.

——— ——— (1950). *Rev.mod.Phys.* 22, 153.

AL'TSHULER, B.L. (1967). *Soviet Phys.JETP* 24, 766.

——— (1969). *Soviet Phys.JETP* 28, 687.

ALY, J.J. (1974). *Astron.Astrophys.* 35, 311.

——— CASER, S., OMNES, R., PUGET, J.I., and VALLADAS, G. (1974). *Astron.Astrophys.* 35, 271.

ANAND, K.C., DANIEL, R.R., and STEPHENS, S.A. (1968). *Phys. Rev.Lett.* 20, 764.

——— ——— ——— (1971). *12th International Conference Cosmic Rays (Hobart).* Vol.7, p.2556.

ANILE, A.M., DANESE, I., ZOTTI, G.DE, and MOTTA, S. (1976). *Astrophys.J.* 205, L59.

ARP, H. (1970). *Nature,Lond.* 22 , 1033.

——— (1971). *Astrophys.Lett.* 7, 221.

——— (1974). *IAU Symposium* 58, p.199-224. Reidel, Dordrecht.

AUSTIN, T.B. and PEACH, J.V. (1974). *Mon.Not.R.astr.Soc.* 167, 437.

BAHCALL, J. (1964). *Phys.Rev.* 136, 1164.

——— and FRAUTSCHI, S. (1971). *Astrophys.J.* 170, L81.

——— and KOZLOVSKY, B.Z. (1969a). *Astrophys.J.* 155, 1077.

BAHCALL, J. ——— (1969b). *Astrophys.J.* 158, 529.

——— and MAY, R.M. (1968). *Astrophys.J.* 152, 37.

——— and OKE, J.B.(1971). *Astrophys.J.* 163, 235.

BALDWIN, J.R., DANZIGER, I.J., FROGEL, J.A., and PERSSON, S.E. (1973). *Astrophys.Lett.* 14, 1.

BALKOWSKI, C., BOTTINELLI, L., CHAMARAUX, P., GOUGIENHEIM, L., and HEIDMANN, J. (1974). *IAU Symposium* 58, 237. Reidel, Dordrecht.

BALLY, J. and HARRISON, E.R. (1978). *Astrophys.J.* 220, 743.

BANDYOPADHYAY, N. (1977). *J.Phys.* A10, 189.

——— (1978a). *J.math.Phys.*, 19, 1423.

——— (1978b). *J.math.Phys.*, in press.

BANERJEE, A. (1967). *Proc.phys.Soc.(Lond)*. 91, 794.

——— (1970). *J.math.Phys.* 11, 51.

——— (1972). *J.Phys.* A5, 1305.

BANERJI, S. (1968). *Prog.theor.Phys., Osaka* 39, 365.

——— (1974). *Phys.Rev.* D9, 877.

BARNOTHY, J.M. and TINSLEY, B.M. (1973). *Astrophys.J.* 182, 343.

BARROW, G.W. (1974). *Mon.Not.R.Astr.Soc.* 175, 359.

BARRY, G.W. (1974). *Astrophys.J.* 190, 279.

BATAKIS, N. and COHEN, J.M. (1975). *Phys.Rev.* D12, 1544.

BAUM, W.A. (1976). *Astrophys.J.* 209, 319.

BECK, F., HILF, E., and MAIR, K. (1 973). *Acta.Phys.Austr.* 38, 201.

BECKMAN, J.E., ADE, P.A.R., HUIZINGA, J.S., ROBSON, E.J., VICKERS, D.G., and HARRIES, J.E. (1972). *Nature,Lond.* 237, 154.

BEERY, J.G., MARTIZ, T.Z., NOLT, I.G., and WOOD, C.W. (1971). *Nature,Lond. (Phys.Sci.)* 230, 36.

BEHR, C. (1962). *Z.Astrophys.* 54, 268.

——— (1965). *Z.Astrophys.* 60, 286.

BEKENSTEIN, J.D.(1975). *Phys.Rev.* D11, 2072.

BELINSKI, V.A. and KHALATNIKOV, I.M. (1969). *Soviet Phys. JETP* 29, 911.

——— ——— and LIFSHITZ, E.M. (1970). *Adv.Phys.* 19, 525.

BELINSKI, V.A., KHALATNIKOV, I.M., and LIFSHITZ, E.M. (1971). *Soviet Phys.JETP* 33, 1061.

——— ——— ——— (1972). *Soviet Phys.JETP* 35, 838.

BENNETT, A.S. (1962). *Mem.R.astr.Soc.* 67, 163.

BERGE, G.L. and SEIELSTAD, G.A. (1967). *Astrophys.J.* 148, 367.

BERTOTTI, B. (1966). *Proc.R.Soc.* A294, 195.

BISHOP, N.T. (1976). *Mon.Not.R.astr.Soc.* 176, 241.

BLACK, J.H. and DALGARNO, A. (1973). *Astrophys.J.* 184, L101.

BLAIR, A.G., BEERY, J.G., EDESKUTZ, F., HIEBERT, R.D., SHIPLEY, J.P., and WILLIAMSON, Jr. K.D. (1971). *Phys. Rev.Lett.* 27, 1154.

BLAKE, G.M. (1976). *Mon.Not.R.astr.Soc.* 174, 63P.

BLEEKER, J.A.M., BURGER, J.J., DEBRENBERG, A.J.M., SCHEEPMAKER, A., SWANENBERG, B.N., and TANAKA, Y. (1968). *Can.J.Phys.* 46, 522.

BLUDMAN, S.A. (1976). *Gen.Relativ.Grav.* 7, 569

BOGDANOVA, L.N. and SHAPIRO, I.S. (1974). *Soviet Phys. JETP* 20, 94.

BOLTON, J.G., GARDNER, F., and MACKAY, M.B. (1964). *Aust. J.Phys.* 17, 340.

BONDI, H. (1947). *Mon.Not.R.astr.Soc.* 107, 410.

——— (1962a). *Cosmology*, 2nd.edn. Cambridge University Press.

——— (1962b). *Observatory* 82, 133.

——— (1969). *Mon.Not.R.astr.Soc.* 142, 333.

——— and GOLD, T. (1948). *Mon.Not.R.astr.Soc.* 108, 252.

——— and LYTTLETON, R.A. (1959). *Proc.R.Soc.* A252, 313.

BONNOR, W.B. (1957). *Mon.Not.R.astr.Soc.* 117, 104.

——— (1967). *Lectures Applied Mathematics (American Mathematical Society)*, Vol.8, p.263.

——— (1972). *Mon.Not.R.astr.Soc.* 159, 261.

——— (1974). *Mon.Not.R.astr.Soc.* 167, 55.

BONOMETTO, S.A., DANESE, L., and LUCCHIN, F. (1974). *Astron. Astrophys.* 35, 267.

——— ——— ——— (1975). *Astron. Astrophys.* 41, 55.

——— LUCCHIN, F., and PUGET, J.L. (1974). *Astron.Astrophys.* 37, 27.

BORTOLOT, Jr., V.J., CLAUSER, J.F., and THADDEUS, P. (1969). *Phys.Rev.Lett.* 22, 307.

BOTTINELLI, L. and GOUGIENHEIM, L. (1973). *Astron.Astrophys.* 26, 85.

BOWLES, J.H., PATRICK, T.J., SHEATHER, P.H., and EIBAND, A.M. (1974). *J.Phys.* E 7, 183.

BOYNTON, P. and PARTRIDGE, R.B. (1973). *Astrophys.J.* 181, 243.

——— and STOKES, R.A. (1974). *Nature,Lond.* 247, 528.

——— and WILKINSON, D.T. (1968). *Phys.Rev.Lett.* 21, 462.

BRACCESI, A., CECCARELLI, M., FANTI, R., and GIOVANNINI, C. (1966). *Nuovo Cim.* B41, 92.

——— ——— ——— GELATO, G., GIOVANNINI, C., HARRIS, D., ROSATELLI, C., SINIGAGLIA, G. and VOLDERS, L. (1965). *Nuovo Cim.* B40, 267.

BRANCH, D. and PATCHETT, B. (1973). *Mon.Not.R.astr.Soc.* 161, 71.

BRANS, C.H. (1962). *Phys.Rev.* 125, 2194.

——— (1975). *Astrophys.J.* 197, 1.

——— and DICKE, R.H. (1961). *Phys.Rev.* 124, 925.

BRECHER, K. (1973). *Astrophys.J.* 181, 255.

——— and BURBIDGE, G.R. (1972). *Nature,Lond.* 237, 440.

——— and MORRISON, P. (1969). *Phys.Rev.Lett.* 23, 802.

——— and SILK, J. (1969). *Astrophys.J.* 158, 91.

BRIDLE, A.H. and DAVIS, M. (1972). *IAU Symposium* 44, p.437. Reidel, Dordrecht.

BRILL, D. (1972). *Evidence for gravitational theories.* Academic Press, London.

BROWN, G.S. and TINSLEY, B.M. (1974). *Astrophys.J.* 194, 555.

BURBIDGE, E.M., BURBIDGE, G.R., FOWLER, W.A., and HOYLE, F. (1957). *Rev.Mod.Phys.* 29, 547.

BURBIDGE, G.R. (1975). *Astrophys.J.* 196, L7.

——— and BURBIDGE, E.M. (1967). *Quasi-stellar objects.* Freeman, New York.

BURKE, J.A. and HARTWICK, F.D.A. (1974). *Astron.Astrophys.* 34, 445.

CADERNI, N., de COSMO, V., FABBRI, R., MELCHIORRI, B., MELCHIORRI, F., and NATALE, V. (1977). *Phys.Rev.* D16, 2424.

CAMERON, A.G.W. (1973). Conference on explosive nucleosynthesis, Austin, Texas. (unpublished)

CAMPUSSANO, L., HEIDEMANN, J., and NIETO, J.L. (1975). *Astron.Astrophys.* 41, 229.

CARLITZ, R., FRAUTSCHI, S., and NAHM, W. (1973). *Astron Astrophys.* 26, 171.

CARR, B.J. (1975). *Astrophys.J.* 201, 76.

CARSWELL, R.F. and STRITTMATTER, P.A. (1973). *Nature,Lond.* 242, 394.

CARTAN, E. (1922). *C.r.hebd.séanc.Acad.Sci.* 174, 593.

CESARSKY, D.A., MOFFET, A.T., and PASACHOFF, J.M. (1973). *Astrophys.J.* 180, L1.

CHANDRASEKHAR, S. and WRIGHT, J.P. (1961). *Proc.natn.Acad. Sci.U.S.A.* 47, 341.

CHAPMAN, G.A. (1975). *Phys.Rev.Lett.* 34, 755.

——— and INGERSOLL, A.P. (1972). *Astrophys.J.* 175, 819.

——— ——— (1973). *Astrophys.J.* 183, 1005.

CHEN, C. and STOTHERS, R. (1976). *Phys.Rev.Lett.* 36, 833.

CHERNIN, A.D.(1965). *Astr.Zh.* 42, 1124.

——— (1968). *Nature,Lond.* 220, 250.

——— (1970). *Nature,Lond.* 226, 440.

——— (1971). *Astrophys.Lett.* 8, 31.

——— (1972). *Astrophys.Lett.* 10, 1.

CHIBISOV, G.V. and OZERNOY, L.M. (1969). *Astrophys.Lett.* 3, 189.

CHITRE, D. (1973). *Phys.Rev.* D6, 3390.

CHRISTENSEN, C.J., NIELSEN, A., BAHNSEN, A., BROWN, W.K., and RUSTAD, B.N. (1967). *Phys.Lett.* B26, 11.

CHU, W.T., KIM, Y.S., BEAM, W.J., and KWAK, N. (1970). *Phys. Rev.Lett.* 24, 917.

CHURCHWELL, E. and MEZGER, P.G. (1973). *Nature,Lond.* 242, 319.

——— ——— and HUCHTMEIER, C. (1974). *Astron.Astrophys.* 32, 2.

CLAUSER, J.F. and THADDEUS, P. (1969). In *Topics in relativistic astrophys* (eds.S.P.Moran and A.G.W. Cameron). Gordon & Breach, New York.

CLEMENCE, G.M. (1957). *Rev.mod.Phys.* 29, 2.

CLINE, D., McINTYRE, P., and RUBBIA, C. (1977). *Phys.Lett.* B66, 429.

COCCONI, G. and SALPETER, E.E. (1958). *Nuovo Cim.* 10, 646.

COHEN, R.L. and WERTHEIM, G.K. (1973). *Nature,Lond.* 241, 109.

COLGATE, S.A. (1973). *Astrophys.J.* 181, L53.

COLLINS, G.B. and HAWKING, S. (1973a). *Astrophys.J.* 180, 317.

——— ——— (1973b). *Mon.Not.R.astr.Soc.* 162, 307.

——— and STEWART, J.M. (1971). *Mon.Not.R.astr.Soc.* 153, 419.

COLLIN-SOUFFRIN, S., PECKER, J.C., and TOVMASSION, H.M. (1974). *Astron.Astrophys.* 30, 351.

CONKLIN, E.K. (1969). *Nature,Lond.* 222, 971.

——— and BRACEWELL, R.N. (1967a). *Nature,Lond.* 216, 777.

——— ——— (1967b). *Phys.Rev.Lett.* 18, 614.

COREY, B.E. and WILKINSON, D.T. (1976). *Bull.Am.astr.Soc.* 8, 351.

COWAN, C.L. (1968). *Astrophys. J.* 154, L5.

COWSIK, R. and KOBETICH, J.E. (1972). *Astrophys.J.* 177, 585.

——— and MCLELLAND, J. (1973). *Astrophys. J.* 180, 7.

———, PAL, Y., and TANDON, S.N. (1964). *Phys.Rev.Lett.* 13, 265.

CRANE, P. and HOFFMAN, A. (1973). *Astrophys.J.* 186, 787.

DALLAPORTA, N. and LUCCHIN, F. (1972). *Astron.Astrophys.* 19, 123.

——— ——— (1973). *Astron.Astrophys.* 26, 325.

DANJO, A., HAYAKAWA, S., MAKINO, F., and TANAKA, Y. (1968). *Can.J.Phys.* 46, 530.

DAS, A. and AGARWAL, P. (1974). *Gen.Relativ.Grav.* 5, 359.

DAUTCOURT, G. (1969). *Mon.Not.R.astr.Soc.* 144, 255.

——— (1975). *Astron.Astrophys.* 38, 335.

DAVIDSON, W. and EVANS, A.B. (1971). *Nature,Lond. (Phys.Sci.)* 232, 29.

DAVIS, M. (1971). *Astr.J.* 76, 980.

——— and WILKINSON, D.T. (1974). *Astrophys.J.* 192, 251.

DAVIS, Jr. R., HARMER, D.S., and HOFFMAN, K.C. (1968). *Phys. Rev.Lett.* 20, 1205.

——— (1972). *Bull.Am.phys.Soc.* 17, 527.

DAY, G.A., SHIMMINS, A.J., EKERS, R.D., and COLE, D.J. (1966). *Aust.J.Phys.* 19, 35.

DE, U.K. (1969). *J.Phys.A* 2, 427.

DEARBORN, D.S. and SCHRAMM, D.N. (1974). *Nature,Lond.* 247, 441.

DEMARET, J. and VAN DER MEULEN, J. (1978). *Phys.Lett.* B73, 471.

DENNIS, B.R., SURI, A.N., and FROST, K.J. (1973). *Astrophys. J.* 186, 97.

DICKE, R.H. (1961). *Nature,Lond.* 192, 440.

——— (1962a). *Phys.Rev.* 125, 2163.

——— (1962b). *Evidence for gravitational theories* Academic Press, London.

——— (1964). *Theoretical significance of experimental relativity*. Blackie, Glasgow.

DICKE, R.H. (1968a). *Astrophys.J.* 152, 1.

——— (1968b). *Astrophys.J.* 154, 892.

——— (1973). *Astrophys.J.* 180, 293.

——— and GOLDENBERG, H.M. (1967). *Phys.Rev.Lett.* 18, 313.

———, PEEBLES, P.J.E., ROLL, P.G., and WILKINSON, D.T. (1965). *Astrophys.J.* 142, 114.

DIRAC, P.A.M. (1938). *Proc.R.Soc.* A165, 199.

——— (1974). *Proc.R.Soc.* A338, 439.

DIXON, M.E. (1966). *Mon.Not.R.astr.Soc.* 131, 325.

DODD, R.J., MORGAN, D.H., NANDY, K., REDDISH, V.C., and SEDDON, H. (1975). *Mon.Not.R.astr.Soc.* 171, 329.

DOROSHKEVICH, A.G., LUKASH, V.N., and NOVIKOV, I.D. (1971). *Soviet Phys.JETP* 33, 649.

——— ——— ——— (1973). *Soviet Phys.JETP* 37, 739.

——— ZEL'DOVICH, YA.B., and NOVIKOV, I.D. (1968). *Soviet Phys.JETP* 26, 408.

——— ——— ——— (1971). *Soviet Phys.JETP* 33, 1.

DURNEY, B.R. and WERNER, N.E. (1971). *Sol.Phys.* 21, 21.

DYER, C.C. and ROEDER, R.C. (1972). *Astrophys.J.* 174, L115.

——— ——— (1973). *Astrophys.J.* 180, L31.

EARDLEY, D., LIANG, E., and SACHS, R. (1972). *J.Math.Phys.* 13, 99.

EDDINGTON, A.S. (1933). *Expanding universe* Cambridge University Press.

——— (1939). *Sci.Prog.,Lond.* 34, 225.

EGGEN, O.J., LYNDEN-BELL, D., and SANDAGE, A.R. (1962). *Astrophys.J.* 136, 748.

EHLERS, J. (1961). *Akad.Wiss.Lit.Mainz.Abh.math.Nat.Kl.* 11, 793.

———, GEREN, P., and SACHS, R.K. (1968). *J.Math.Phys.* 9, 1344.

——— (1971). *Relativity and cosmology*. Academic Press, London.

EINASTO, J., KAASIK, A., and SAAR, E. (1974). *Nature,Lond.* 250, 309.

——— SAAR, E., KAASIK, A., and CHERNIN, A.D. (1974). *Nature,Lond.* 252, 111.

EINSTEIN, A. (1950). *The meaning of relativity*. Methuen, London.

——— and DE SITTER, W. (1932). *Proc.natn.Acad.Sci. U.S.A.* 18, 213.

EINSTEIN, A. and STRAUSS, E.G. (1945). *Rev.mod.Phys.* 17, 120.

EISENHART, L.P. (1933). *Continuous groups of transformations.* Princeton University Press, N.J.

ELLIS, G.F.R. (1971). *Relativity and cosmology.* Academic Press, London.

——— and KING, A.R. (1974). *Communs.Math.Phys.* 38, 119.

and MACCALLUM, M.A.H. (1969). *Communs.Math.Phys.* 12, 108.

ELLIS, R.S., FONG, F., and PHILLIPS, S. (1976). *Mon.Not.R. astr.Soc.* 176, 391.

EPSTEIN, R.L. and PETROSIAN, V. (1975). *Astrophys.J.* 197, 281.

ESTABROOK, F.B., WAHLQUIST, H.D., and BEHR, C.G. (1968). *J. Math.Phys.* 9, 497.

EULER, H. and KOCKEL, B. (1935). *Naturwissenschaften* 23, 346.

EVANS, A.B. (1975). *Phys.Lett.* A55, 271.

——— and DAVIDSON, W. (1973). *Mon.Not.R.astr.Soc.* 165, 323.

EWING, M.S., BURKE, B.F., and STAELIN, D.H. (1967). *Phys. Rev.Lett.* 19, 1251.

FANAROFF, B.L. and LONGAIR, M.S. (1973). *Mon.Not.R.astr.Soc.* 161, 393.

FARNSWORTH, D.L. (1967). *J.Math.Phys.* 8, 2315.

FELTON, J.E. and MORRISON, P. (1966). *Astrophys.J.* 146, 686.

FENNELLY, A.J. (1976). *Astrophys.J.* 207, 693.

FIELD, G.B. (1959). *Astrophys.J.* 129, 525.

——— (1969). *Astrophysics and general relativity*, vol. 1. Gordon and Breach, London.

——— and SHEPLEY, L.C. (1968). *Astrophys.Space.Sci.* 1, 309.

———, ARP, H., and BAHCALL, J.N. (1973). *The red shift controversy.* Benjamin, New York.

FITCH, L.T., DIXON, R.S., and KRAUS, J.D. (1969). *Astr.J.* 74, 612.

FOCK, V. (1935). *Z.Physik.* 98, 148.

FOMALONT, E.B., BRIDLE, A.H., and DAVIS, M.M. (1974). *Astron. Astrophys.* 36, 273.

FOWLER, W.A. and HOYLE, F. (1973). *Nature,Lond.* 241, 384.

FRAUTSCHI, S., STEIGMAN, G., and BAHCALL, J.N. (1972). *Astrophys.J.* 175, 307.

FRIEDMANN, A. (1922). *Z.Phys.* 10, 377.

FUJIMOTO, M., KAWABATA, K., and OSFUE, Y. (1971). *Prog.Theor. Phys., Osaka* Suppl. 49, 181.

GALTON, J.A. and KENNEDY, J.E.D. (1968). *Astr.J.* 73, 135.

GANGULI, S.N., MALHOTRA, P.K., RAGHARAN, R., SUBRAMANIAN, A. and SUDHAKAR, K. (1978). *Phys.Lett* . 74B, 130.

GARDNER, F.F., and WHITEOAK, J.B. (1969).*Austr.J.Phys.* 22, 107.

——— MORRIS, D., and WHITEOAK, J.B. (1969). *Austr. J.Phys.* 22, 79.

GELLER, M.J. and PEEBLES, P.J.E. (1972). *Astrophys.J.* 174, 1.

GEROCH, R. (1967). Ph.D.thesis. Princeton University, N.J.

——— (1970a). *J.Math.Phys.* 11, 437.

——— (1970b). *Relativity* (eds. M.Carmeli, *et al*) Plenum Press, New York.

GILMAN, R.C. (1970). *Phys.Rev.* D2, 1400.

GINZBURG, V.L. (1971). *Soviet Phys. Usp.* 14, 21.

——— and OZERNOY, L.M. (1966). *Soviet Phys.* 9, 726.

GISLER, G.R., HARRISON, E.H., and REES, M.J. (1974). *Mon.Not. R.astr.Soc.* 166, 663.

GÖDEL, K. (1949). *Rev.mod.Phys.* 21, 447.

——— (1950). *Proc.Int.Cong.Math.* (Cambridge, Mass). 1, 175.

GOLD, T. and PACINI, F. (1968). *Astrophys.J.* 152, L115.

GOLDEN, L.M. (1974). *Mon.Nat.R.astron.Soc.* 166, 383.

GOLDONI, R. (1976a). *Gen.Relativ.Grav.* 7, 731.

——— (1976b). *Gen.Relativ.Grav.* 7, 743.

GORENSTEIN, P., BJORKHOLM, P., HARRIS, B., and HARNDEN, Jr. F. (1973). *Astrophys.J.* 183, L57.

GOTT, III, J.R., GUNN, J.E., SCHRAMM, D.N., and TINSLEY, B.M. (1974). *Astrophys.J.* 194, 543.

GOULD, R.J. and SCHREDER, G.P. (1966). *Phys.Rev.Lett.* 16, 252.

——— ——— (1967). *Phys.Rev.* 155, 1408.

GOWER, J.F.R. (1966). *Mon.Not.R.astr.Soc.* 133, 151.

———, SCOTT, P.F., and WILLS, D. (1967). *Mem.R.astr. Soc.* 71, 49.

DE GRAFF, T. (1970). *Lett.Nuovo Cim.* 4, 638.

GREENSTEIN, G.S. (1968a). *Astrophys.Lett.* 1, 139.

——— (1968b). *Space Sci.* 2, 155.

GREENSTEIN, J. (1966). *Astrophys.J.* 144, 496.

GREISEN, K. (1966). *Phys.Rev.Lett.* 16, 748.

——— (1971). *Astrophysics and general relativity*, Vol.2. Gordon and Breach, London.

GRENIER, P., ROUCHER, J., and TALUREAU, B. (1976). *Astron. Astrophys.* 53, 249.

GRIFFITHS, R.E. and PEACOCK, A. (1974). *Nature,Lond.* 250, 471.

GRISCHUK, L.P., DOROSHKEVICH, A.G., and NOVIKOV, I.D. (1969). *Soviet Phys. JETP* 28, 1210.

GROTH, E.G. and PEEBLES, P.J.E. (1975). *Astron.Astrophys.* 41, 143.

GRUEFF, G. and VIGOTTI, M. (1968). *Astrophys.Lett.* 2, 113.

GULL, S.F. and NORTHOVER, K.J.E. (1976). *Nature,Lond.* 263, 572.

GUNN, J.E. (1967). *Astrophys.J.* 147, 61.

——— and GOTT, III, J.R. (1972). *Astrophys.J.* 176, 1.

——— and OKE, J.B. (1975). *Astrophys.J.* 195, 255.

——— and PETERSON, B.A. (1965). *Astrophys.J.* 142, 1633.

——— and TINSLEY, B.M. (1975). *Nature,Lond.* 257, 454.

——— ——— (1976). *Astrophys.J.* 210, 1.

GURSEY, F. (1963). *Ann.Phys.* 24, 211.

GURSKY, H., SOLINGER, A., KELLOG, E.M., MURRAY, S., TANANBAUM, H., GIACCONI, R., and CAVALIERE, A. (1972). *Astrophys.J.* 173, L99.

GUTH, E. (1969). *Phys.Rev.Lett.* 21, 106.

HAGEDORN, R. (1965). *Nuovo Cim.Suppl.* 3, 147.

——— (1970). *Astron.Astrophys.* 5, 184.

HAGGERTY, M.J. and WERTZ, J.R. (1972). *Mon.Not.R.astr.Soc.* 155, 495.

HAINEBACH, K.L. and SCHRAMM, D.N. (1976). *Astrophys.J.* 207, L79.

HAMER, C., and FRAUTSCHI, S. (1971). *Phys.Rev.* D4, 2125.

HARRIS, B.J. and KRAUS, J.D. (1970). *Nature,Lond.* 227, 785.

——— ——— (1971). *Nature,Lond. (Phys.Sci.)* 230, 140.

HARRISON, E.R. (1964). *Nature,Lond.* 204, 271.

——— (1965). *Mon.Not.R.astr.Soc.* 131, 1.

——— (1967). *Rev.mod.Phys.* 39, 862.

——— (1968). *Phys.Rev.* 167, 1170.

——— (1969). *Nature,Lond.* 224, 1089.

——— (1970a). *Mon.Not.R.astr.Soc.* 147, 279.

——— (1970b). *Nature,Lond.* 228, 258.

——— (1972a). *Phys.Rev.* D6, 2077.

——— (1972b). *Comments Astrophys.Space Phys.* 4, 187.

HARRISON, E.R. (1975). *Astrophys.J.* 195, L61.

HAWKING, S. (1966). *Astrophys.J.* 145, 544.

——— (1967). *Proc.R.Soc.* A300, 187.

——— (1969). *Mon.Not.R.astr.Soc.* 142, 129.

——— (1974). *Nature,Lond.* 248, 30.

——— and ELLIS, G.F.R. (1968). *Astrophys.J.* 152, 25.

——— ——— (1973). *Large scale structure of spacetime.* Cambridge University Press.

——— and PENROSE, R. (1970). *Proc.R.Soc.* A314, 529.

——— and TAYLER, R.J. (1966). *Nature,Lond.* 209, 1278.

HAWKINS, M.R.S. and MARTIN, R. (1977). *Nature,Lond.* 265, 711.

HAYASHI, K. (1950). *Prog.theor.Phys., Osaka* 5, 224.

——— and SHIRAFUJI, T. (1977). *Prog.theor.Phys., Osaka* 57, 302.

HECHT, H.F. (1971). *Astrophys.J.* 170, 401.

——— (1973). *Astron.Astrophys.* 26, 123.

HECKMANN, O. (1961). *Astr.J.* 66, 599.

——— (1968). *Theorien der Kosmologie.* Springer Verlag, Berlin.

——— and SCHUCKING, E. (1955). *Z.Astrophys.* 38, 95.

——— ——— (1959). *Handb.Phys.* 53, 489.

——— ——— (1962). *Gravitation.* Wiley, New York.

HEGYI, D.J., TRAUB, W.A., and CARLETON, N.P. (1974). *Astrophys.J.* 190, 543.

HEHL, F.W. (1966). *Abh.braunschw.wiss.Ges.* 18, 98.

——— (1973). *Gen.Relativ.Grav.* 4, 333.

——— (1974). *Gen.Relativ.Grav.* 5, 491.

——— and VON DER HEYDE, P. (1973). *Annls.Inst.Henri Poincaré,* A19, 179.

——— ——— and KERLICK, G.D. (1974). *Phys.Rev.* D10, 1066.

——— ——— ——— and NESTER, J.M. (1976). *Rev. mod. Phys.* 48, 393.

HEINTZMAN, H. and HILLERBRANDT, W. (1975). *Phys.Lett.* A54, 349.

HELLER, M. (1970a). *Acta phys.pol.* B1, 123.

——— (1970b). *ACta phys.pol.* B1, 131.

——— and SUSZYCKI, L. (1974). *Acta phys.pol.* B5, 345.

——— KLIMEK, Z., and SUSZYCKI, L. (1973). *Astrophys. Space Sci.* 20, 205.

HENRY, P.S. (1971). *Nature,Lond.* 231, 516.

HEWISH, A. (1961). *Mon.Not.R.astr.Soc.* 123, 167.

HILL, H.A. and STEBBINS, R.T. (1975). *Astrophys.J.* 200, 471.

HINDER, R.A. and BRANSON, N.J.B.A. (1969). *Observatory* 89, 178.

VAN HOERNER, S. (1973). *Astrophys.J.* 186, 741.

HOLDEN, D.J. (1966). *Mon.Not.R.astr.Soc.* 133, 225.

HOOFT, G'T. (1971). *Phys.Lett.* B37, 197.

HOPPER, V.D., MACE, O.B., THOMAS, J.A., ABBOTS, P., FRYE, JR. G.M., THOMSON, G.B., and STAIB, J.A. (1973). *Astrophys.J.* 186, L55.

HOUCK, J.R. and HARWIT, M.O. (1969). *Astrophys.J.* 157, L45.

——— , SOIFER, B.T., HARWIT, M.O., and PIPHER, J.L. (1972). *Astrophys.J.* 178, L29.

HOWELL, T.F. and SHAKESHAFT, J.R. (1966). *Nature,Lond.* 210, 1318.

——— ——— (1967). *Nature,Lond.* 216, 753.

HOYLE, F. (1948). *Mon.Not.R.astr.Soc.* 108, 372.

——— (1954). *Astrophys.J.* Suppl., Ser 1, No.5121.

——— (1960). *Proc.R.Soc.* A257, 431.

——— (1962). *Evidence for gravitational theories.* Academic Press, London.

——— (1975). *Astrophys.J.* 196, 661.

——— and NARLIKAR, J.V. (1963). *Proc.R.Soc.* A273, 1.

——— ——— (1964). *Proc.R.Soc.* A278, 465.

——— ——— (1970). *Nature,Lond.* 228, 544.

——— ——— (1971). *Nature,Lond.* 233, 41.

——— ——— (1972a). *Mon.Not.R.astr.Soc.* 155, 305.

——— ——— (1972b). *Mon.Not.R.astr.Soc.* 155, 323.

——— and WICKRAMSINGHE, N.C. (1967). *Nature,Lond.* 214, 969.

HUANG, K. and WEINBERG, S. (1970). *Phys.Rev.Lett.* 25, 895.

HUCHTMEIER, W.K. and BATCHELOR, R.A. (1973). *Nature,Lond.* 243, 154.

HUGHES, R.G. and LONGAIR, M.S. (1967). *Mon.Not.R.astr.Soc.* 135, 131.

HUGHES, V.W., ROBINSON, H.G., and BELTROW-LOPEZ, V. (1960). *Phys.Rev.Lett.* 4, 342.

HUNTER, C. (1964). *Astrophys.J.* 139, 570.

INGERSOLL, A.P. and SPIEGEL, E.A. (1971). *Astrophys.J.* 163, 375.

IRVINE, N. (1965). *Ann.Phys.* 30, 322.

ISAACSON, R.A. (1968). *Phys.Rev.* 166, 1272.

——— and WINICOUR, J. (1973). *Astrophys.J.* 184, 49.

ISHAM, C.J., SALAM, A. and STRATHDEE, J. (1971). *Phys.Rev.* D.3, 867.

——— ——— ——— (1973). *Nature, Lond., Phys. Sci.* 244, 82.

JAAKKOLA, T. (1971). *Nature,Lond.* 234, 534.

——— (1973). *Astron.Astrophys.* 27, 449.

——— and LE DENMOT, G. (1976). *Mon.Not.R.astr.Soc.* 176, 307.

JACOBS, K.C. (1968). *Astrophys.J.* 153, 661.

——— (1969). *Astrophys.J.* 155, 379.

JEANS, J. (1929). *Astronomy and cosmogony* . Cambridge University Press.

JEFFERTS, K.B., PENZIAS, A.A., and WILSON, R.W. (1973). *Astrophys.J.* 179, L57.

JELLEY, J.V. (1966). *Phys.Rev.Lett.* 16, 479.

JOHRI, V.B. (1972). *Tensor* 25, 241.

JONES, B.F. (1976). *Astr.J.* 81, 455.

JONES, B.J.T. (1973). *Astrophys.J.* 181, 269.

——— (1976). *Rev.mod.Phys.* 48, 107.

JORDAN, P. (1955). *Scherkraft und Weltall.* Vieweg, Braunschweig.

JOSEPH, V. (1957). *Proc.Camb.phil.Soc.math.phys.Sci.* 53, 836.

JURA, M. (1973). *Astrophys.J.* 181, 627.

KAFKA, P. (1970). *Nature,Lond.* 226, 436.

KANTOWSKI, R. and SACHS, R.K. (1966). *J.Math.Phys.* 7, 443.

KAPAHI, V.K. (1975). *Mon.Not.R.astr.Soc.* 172, 513.

KARDASHEV, N. (1967). *Astrophys.J.* 150, L135.

KASNER, E. (1927). *Am.J.Math.* 43, 217.

KASTURIRANGAM, K. and RAO, U.R. (1972). *Astrophys.Space Sci.* 15, 161.

KATGERT, P., KATGERT-MERKELIJN, J.K., LEPOOLE, R.S., and VAN DER LAAR, H. (1973). *Astron.Astrophys.* 23, 171.

KAUFMAN, S.E. (1971). *Astr.J.* 76, 751.

——— (1973). *Nuovo Cim.* B13, 91.

——— and SCHUCKING, E.L. (1971). *Astr.J.* 76, 583.

KELLERMAN, K.I., PAULINY-TOTH, I.I.K., and DAVIS, M.M. (1968). *Astrophys.Lett.* 2, 113.

——— ——— and WILLIAM, P.J.S. (1969). *Astrophys.J.* 157, 1.

KEMBHAVI, A.K. (1976). *Pramana, India* 7, 344.

KERLICK, G.D. (1975a). Ph.D.thesis, Princeton University.

——— (1975b). *Phys.Rev.* D12, 3004.

——— (1976). *Ann.Phys. N.Y.* 99, 127.

———, KIBBLE, T.W.B. (1961). *J.math.Phys.* 2, 212.

KING, A.R. (1974). *Commun.Math.Phys.* 38, 117.

——— and ELLIS, G.F.R. (1973). *Commun.Math.Phys.* 31, 209.

KING, J.G. (1960). *Phys.Rev.Lett.* 5, 562.

KIRSHNER, R.P. and KWAN, J. (1974). *Astrophys.J.* 193, 27

KISLYAKOV, A.G., CHERNYSHEV, V.I., LEBSKII, YU.V., MAL'TSEV, V.A., and SEROV, N.V. (1971). *Soviet Astr.* 15, 29.

KLEIN, O. (1958). *La structure et l'evolution de l'universe.* Stoops.

KOEHLER, J.A. and ROBINSON, B.J. (1966). *Astrophys.J.* 146, 488.

KOHLER, M. (1933). *Annl.Phys.* 16, 129.

KOMPANIETS, A.S. and CHERNOV, A.S. (1969). *Soviet Phys. JETP* 20, 1303.

KONSTANTINOV, B.P. and KOCHAROV, G.E. (1964). *Soviet Phys. JETP* 19, 992.

——— ———, and STARBUNOV, YU.N. (1968). *Izvestia* 32, 1841.

KOPCZYNSKI, W. (1972). *Phys.Lett.* A39, 219.

——— (1973). *Phys.Lett.* A43, 63.

KRISTIAN, J. and SACHS, R.K. (1966). *Astrophys.J.* 143, 379.

KRUSZEWSKI, A. and SEMENIUK, I. (1976). *Acta astr.Pol.* 26, 193.

KUCHOWICZ, B. (1973). University of Warsaw preprint UW-R/73/2.

——— (1975a). *Phys.Lett.* A54, 13.

——— (1975b). *Acta phys.Pol* B6, 173.

——— (1975c). *Curr.Sci.* 44, 537.

——— (1976a). *Acta phys.Pol.* B7, 81.

——— (1976b). *J.Phys.* A8, L29.

——— (1976c). *Rep.Prog.Phys.* 39, 291.

——— (1978). *Gen.Relativ.Grav.* 9, 511.

KUNDT, W. (1971). *Springer Tracts mod.Phys.* 58, 1.

KUO, F., FRYE, JR. G.M., and ZYCH, A.D. (1973). *Astrophys.J.* 186, L51.

LAKE, G. and PARTRIDGE, R.B. (1976). *Bull.Am.astr.Soc.* 8, 354.

LANDAU, L.D. and LIFSHITZ, E.M. (1959). *Fluid mechanics.* Pergamon Press, Oxford.

——— ——— (1971). *Classical theory of fields*. Pergamon Press, Oxford.

LANDSBERG, P.T. and BROWN, B.M. (1973). *Astrophys.J.* 182, 653.

——— and PARK, D. (1975). *Proc.R.Soc.* A346, 485.

LANG, K.R., LORD, S.D., JOHANSON, J.M., and SAVAGE, P.D. (1975). *Astrophys.J.* 202, 583.

LANSBERG, A. (1969). *Astron.Astrophys.* 3, 150.

LARSON, R.B. (1974). *Mon.Not.R.astr.Soc.* 166, 585.

——— and TINSLEY, B. (1974). *Astrophys.J.* 192, 293.

LAURENT, B.E. and SODERHOLM, L. (1969). *Astron.Astrophys.* 3,197.

LAYZER, D. (1954). *Astr.J.* 59, 268.

——— and HIVELY, R. (1973). *Astrophys.J.* 179, 361.

LEE, H., LEUNG, Y.C., and WANG, C.G. (1971). *Astrophys.J.* 166, 387.

LEMAITRE, A. (1931). *Mon.Not.R.astr.Soc.* 91, 490.

LIANG, E. (1972). *J.Math.Phys.* 13, 386.

——— (1973). *Commun.Math.Phys.* 32, 51.

——— (1975). *Phys.Lett.* A51, 141.

——— (1976). *Astrophys.J.* 204, 235.

LIEBES, S. (1964). *Phys.Rev.* 133, 835.

LIFSHITZ, E.M .(1946). *Fiz.Zh.* 10, 116.

——— and KHALATNIKOV, I.M .(1963). *Adv.Phys.* 12, 185.

LONGAIR, M.S. (1971). *Rep.Prog.Phys.* 34, 1125.

——— (1974). *IAU Symposium* 58, p.69. Reidel, Dordrecht

——— and SUNYAEV, R.A. (1969). *Nature,Lond.* 223, 719.

LUKÁCS, B. (1976). *Gen.Relativ.Grav.* 7, 653.

LUKASH, V.N. (1974). *Soviet Phys.JETP Lett* . 19, 267.

LUKE, S.K. and SZAMOSI, G. (1972). *Astron.Astrophys.* 20, 397.

LYNDEN-BELL, D. (1967). *Mon.Not.R.astr.Soc.* 135, 453.

——— (1969). *Nature,Lond.* 223, 690.

——— (1977). *Nature,Lond.* 270, 396.

MACCALLUM, M.A.H. (1971). *Nature,Lond.(Phys.Sci)* 230, 112.

MACHALSKI, J., ZIEBA, S., and MASLOWSKI, J. (1974). *Astron. Astrophys.* 33, 357.

MAITRA, S.C. (1966). *J.Math.Phys.* 7, 1025.

MANSFIELD, V.N. (1976). *Nature,Lond.* 261, 560.

MARCHANT, A. and MANSFIELD, V.N. (1977). *Nature,Lond.* 270, 699.

MAROCHNIK, L.S. PELIKHOV, N.V., and VERESHKOV, G.M. (1975a). *Astrophys.Space Sci.* 34, 249.

——— ——— ——— (1975b). *Astrophys.Space Sci.* 34, 281.

MASLOWSKI, J. (1973). *Acta astr.* 22, 227.

———, MACHALSKI, J., and ZIEBA, S. (1973). *Astron. Astrophys.* 28, 289.

MATHER, J.C., WERNER, M.W., and RICHARDS, P.L. (1971). *Astrophys.J.* 170, L59.

MATSUDA, T., SATO, H., and TAKEDA, H. (1971). *Prog.theor.Phys. Osaka* 46, 416.

MATSUO, M., MIKUNO, E., NISHIMURA, J., NIU, K., and TAIRA, T. (1971). *12th International conference cosmic rays (Hobart).* Vol.7, p.2550.

MATTIG, W. (1958). *Astr.Nachr.* 284, 109.

——— (1959). *Astr.Nachr.* 285, 1.

MATZNER, R.A. (1970). *J.Math.Phys.* 11, 2432.

———, RYAN, M.P., and TOTON, E.T. (1973). *Nuovo Cim.* B14, 161.

———, SHEPLEY, L.C., and WARREN, J.B. (1970). *Ann. Phys. N.Y.* 57, 401.

MCCAMMON, D., BUNNER, A.N., COLEMAN, P.L., and KRAUSHAAR, W.L. (1971). *Astrophys.J.* 168, L33.

MCCREA, W.H. (1951). *Proc.R.Soc.* A206, 562.

——— (1954). *Astr.J.* 60, 271.

——— and MILNE, E.A. (1934a). *Q.Jl.Math.* 5, 73.

——— ——— (1934b). *Q.Jl.Math.* 5, 76.

MCINTOSH, C.B.G. (1967). *Nature,Lond.* 216, 1297.

——— (1973). *Phys.Lett.* A43, 33.

MCKELLAR, A. (1941). *Publs.Dom.astrophys.* Obs. 7, 251.

MCVITTIE, G.C. (1933). *Mon.Not.R.astr.Soc.* 93, 325.

——— (1938). *Observatory* 61, 209.

MEHRA, J. (1974). *Einstein. Hilbert and the theory of gravitation.* Reidel, Dordrecht.

MEIER, D.L. (1976). *Astrophys.J.* 203, L103.

MENEGUZZI, M., AUDOUZE, A., and REEVES, H. (1971). *Astron. Astrophys.* 15, 337.

MESZAROS, P. (1975). *Astron.Astrophys.* 38, 5.

MILEY, G.K., PEROLA, G.C., VAN DER KRUIT, P.C., and VAN DER LAAR, H. (1972). *Nature,Lond.* 237, 269.

MILLEA, M.F., MCCOLL, M., PEDERSON, R.J., and VERNON, JR., F.L. (1971). *Phys.Rev.Letts.* 26, 919.

MILLS, B.Y., DAVIES, I.M., and ROBERTSON, J.G. (1973). *Austr. J.Phys.* 26, 417.

MILNE, A.E. (1935). *Relativity, gravitation, and world structure.* Clarendon Press, Oxford.

MILTON, K.A. and NG, Y.J. (1974). *Phys.Rev.* D10, 420.

MISNER, C.W. (1963). *J.Math.Phys.* 4, 924.

MISNER, C.W. (1968). *Astrophys.J.* 151, 431.
——— (1969a). *Phys.Rev.* 186, 1328.
——— (1969b). *Phys.Rev.Lett.* 22, 1071.
——— (1973). *Phys.Rev.* D8, 3271.
——— and TAUB, A.H. (1969). *Soviet Phys.JETP* 28, 122.

MISRA, R.M. and SRIVASTAVA, D.C. (1973). *Phys.Rev.* D8, 1653.

MITTON, S. (1973). *Mon.Not.R.phys.Soc.* 155, 373.
——— and REINHARDT, M. (1972). *Awtron.Astrophys.* 20, 337.

MORGANSTERN, R.E. (1971a). *Phys.Rev.* D3, 2946.
——— (1971b). *Phys.Rev.* D4, 278.
——— (1971c). *Phys.Rev.* D4, 281.
——— (1972). *Nature,Lond. (Phys.Sci.)* 237, 70.
——— (1973). *Phys.Rev.* D7, 1570.

MORRISON, L.V. (1973). *Nature,Lond.* 241, 519.

MUEHLNER, D. and WEISS, R. (1970). *Phys.Rev.Lett.* 24, 742.
——— ——— (1973a). *Phys.Rev.Lett.* 30, 757.
——— ——— (1973b). *Phys.Rev.* D7, 326.

MUHLEMAN, D.O., EKERS, D.O., and FOMALONT, E.B. (1970). *Phys.Rev.Lett.* 24, 1377.

MULLER, D. and MEYER, P. (1973). *Astrophys.J.* 186, 841.

MURPHY, G.L. (1973). *Phys.Rev.* D8, 4231.

NAHM, W. (1972). *Nucl.Phys.* B45, 525.

NANOS, G.P. (1973). Ph.D.thesis. Princeton, N.J. (Quoted by Brans, C.H. (1975). *Astrophys.J.* 197, 1.)

NARIAI, H. (1972). *Prog.theor.Phys., Osaka* 47, 1824.
——— (1974a). *Prog.theor.Phys., Osaka* 51, 613.
——— (1974b). *Prog.theor.Phys., Osaka* 52, 1539.
——— (1975a). *Prog.theor.Phys., Osaka* 53, 656.
——— (1975b). *Prog.theor.Phys., Osaka* 54, 1356.

NARLIKAR, J.V. (1963). *Mon.Not.R.astr.Soc.* 126, 203.
——— (1973). *Nature,Lond.* 242, 135.
——— (1974). *Pramana, India* 2, 158.
——— (1977a). *Ann.Phys.* 107, 325.
——— (1977b). Inertia and cosmology. In *Einstein centenary volume*. Acad.Naz. dei Lincei, Rome. In Press.
———, and CHITRE, S.M. (1977). *Mon.Not.R.astr.Soc.* 180, 525.
——— and SUDARSHAN, E.C.G. (1976). *Mon.Not.R.astr.Soc.* 175, 105.

NARLIKAR, J.V., and WICKRAMSINGHE, N.C. (1967). *Nature,Lond.* 216, 43.

——— ——— (1968). *Nature,Lond.* 217, 1235.

NELSON, A.H. (1973). *Nature,Lond.* 241, 185.

NEUGEBAUER, G. and STROBEL, H. (1969). *Wiss.Z.Friedrich Schiller-Univ.Jena* 18.

NEUMANN, C. and SEELIGER, H. (1896. *Uber das Newtonsche Prinzip der Fernwirkung.* Leipzig.

NICKERSON, B.G. and PARTRIDGE, R.B. (1971). *AStrophys.J.* 169, 203.

NICOLL, J.F. and SEGAL, I.E. (1975). *Proc.natn.Acad.Sci. . . .* 72, 4691.

NIGHTINGALE, J.D. (1973). *AStrophys.J.* 185, 105.

NOONAN, T.W. (1972). *A Ætrophys.J.* 171, 209.

NOVELLO, M. (1976). *Phys.Lett.* A59, 105.

——— (1977a). Orange Aid Preprint 487. Caltech, California.

——— (1977b). Preprint *Galaxy formation and vacuum quantum fluctuations of the gravitational field.*

NOVIKOV, I.D. and ZEL'DOVICH, YA.B. (1967). *A.Rev.Astr. Astrophys.* 5, 627.

O'HANLON, J. and TUPPER, B.O.J. (1972). *Nuovo Cim.* B7, 305.

OKE, J.B. (1971). *Astrophys.J.* 170, 193.

———, and SANDAGE, A.R. (1968). *Astrophys.J.* 154, 21.

OLSON, D.W. (1976). *Phys.Rev.* D14, 327.

OMNES, R. (1972). *Phys.Rep.* 3C, 1.

——— and PUGET, J.L. (1974). *Confrontation of cosmological theories with observational data.* Reidel, Dordrecht.

OORT, J. (1970a). *Astron.Astrophys.* 7, 381.

——— (1970b). *Astron.Astrophys.* 7, 405.

——— (1970c). *Science* 170, 1363.

OPPENHEIMER, R.J. (1971). *Science and synthesis.* Springer Verlag, Berlin.

OSTRIKER, J.P. and TINSLEY, B.M. (1975). *Astrophys.J.* 201, L51.

——— and TREMAINE, S.D. (1975).*Astrophys.J.* 242, L113.

———, PEEBLES, P.J.E., and YAHIL, A. (1974). *Astrophys.J.* 193, L1.

OZERNOY, L.M. and CHERNIN, A.D. (1968). *Soviet Astr.* 11, 907.

——— ——— (1969). *Soviet Astr.* 12, 901.

—— and CHIBISOV, G.V. (1971). *Astrophys.Lett.* 7, 201.

OZSVATH, I. (1965). *J.Math.Phys.* 6, 590.

—— (1966). Perspectives in general relativity. Indiana University Press, Bloomington, Ind.

—— (1967). *J.Math.Phys.* 8, 326.

—— (1970). *J.Math.Phys.* 11, 2860.

—— (1971). *J.Math.Phys.* 12, 1078.

—— and SCHUCKING, E. (1969). *Ann.Phys.* 55, 166.

PAPAPETROU, A. (1951). *Proc.R.Soc.,Lond.* A209, 248.

PARIJSKIJ, Y.N. (1972). *Soviet Astr.* 16, 1048.

—— (1973a). *Soviet Astr.* 17, 291.

—— (1973b). *Astrophys.J.* 180, L47.

PARKER, L., and FULLING, S.A. (1973). *Phys.Rev.* D7, 2357.

PARTRIDGE, R.B. (1974). *Astrophys.J.* 192, 241.

—— and PEEBLES, P.J.E. (1967a). *Autrophys.J.* 147, 757.

—— —— (1967b). *Astrophys.J.* 148, 377.

—— and WILKINSON, D.T. (1967). *Phys.Rev.Lett.* 18, 557.

PAULI, W. (1958). *Theory of relativity*. Pergamon Press, Oxford.

Pauliny-Toth, I.I.K., and Kellerman, K.I. (1972a). *Astr.J.* 77, 265.

—— —— (1972b). *Astr.J.* 77, 560.

—— —— (1972c). *Astr.J.* 77, 797.

—— —— (1974). *Confrontation of cosmological theories with observational data*, p.111. Reidel, Dordrecht.

—— and SHAKESHAFT, J.R. (1962). *Mon.Not.R. astr.Soc.* 124, 61.

PEARSON, T.J. (1974). *Mon.Not.R.Astr.Soc.* 166, 249.

PECKER, J.C., ROBERTS, A.P., and VIGIER, J.P. (1972). *Nature, Lond.* 237, 227.

——, TAIT, W., and VIGIER, J.P. (1973). *Nature, Lond.* 241, 338.

PEEBLES, P.J.E. (1967). *Astrophys.J.* 147, 859.

—— (1966). *Astrophys.J.* 146, 542.

—— (1968). *Astrophys.J.* 153, 1.

—— (1974a). *Astrophys.J.* 189, L251.

—— (1974b). *Astron.Astrophys.* 32, 391.

PEEBLES, and DICKE, R.H. (1968). *Astrophys.J.* 154, 892.

—— and HAUSER, M.G. (1974). *Astrophys.J.* Suppl., 28, 19.

—— and WU, J.T. (1970). *Astrophys.J.* 162, 815.

PENZIAS, A.A. and SCOTT, E.H. (1968). *Astrophys.J.* 153, L7.

—— —— (1969). *Astrophys.J.* 156, 799.

——, JEFFERTS, K.B., and WILSON, R.W. (1972). *Phys. Rev.Lett.* 28, 772.

——, SCHRAML, J., and WILSON, R.W. (1969). *Astrophys. J.* 157, L49.

PERKO, T.E., MATZNER, R.A., and SHEPLEY, L.C. (1972). *Phys. Rev.* D6, 969.

PETROSIAN, V., SALPETER, E., and SZEKERES, P. (1967). *Astrophys.J.* 147, 1222.

PILKINGTON, J.D.H. and SCOTT, P.F. (1965). *Mem.R.astr.Soc.* 69, 183.

PIPHER, J.L., HOUCK, J.R., JONES, B.W., and HARWIT, M. (1975). *Nature,Lond.* 231, 375.

POCHODA, P. and SCHWARZSCHILD, M. (1964). *Astrophys.J.* 139, 587.

POOLEY, G.G. (1969). *Mon.Not.R.astr.Soc.* 144, 101.

—— and KENDERDINE, S. (1968). *Mon.Not.R.astr.Soc.* 139, 539.

—— and RYLE, M. (1968). *Mon.Not.R.astr.Soc.* 139, 515.

PREDMORE, C.R., GOLDWIRE, H.C., and WALTERS, G.K. (1971). *Astrophys.J.* 168, L125.

PRESS, W.H. and GUNN, J.E. (1973). *.J.* 185, 397.

—— and SCHECHTER, P. (1974). *Astrophys.J.* 187, 425.

PRICE, R.M. and MILNE, D.K. (1965). *Austr.J.Phys.* 18, 329.

PUGET, J.L. and SCHATZMANN, E. (1974). *Astron.Astrophys.* 32, 477.

PUZANOV, V.I., SALOMONOVICH, A.E., and STARKOVICH, K.S. (1967). *Soviet Astr.* 11, 905.

QUIRK, W.J. and TINSLEY, B. (1973). *Astrophys.J.* 179, 69.

RAINE, D.F. (1975). *Mon.Not.R.astr.Soc.* 171, 507.

RASBAND, S.N. (1971). *Astrophys.J.* 170, 1.

RAYCHAUDHURI, A.K. (1952). *Phys.Rev.* 86, 90.

—— (1955a). *Phys.Rev.* 98, 1123.

—— (1955b). *Z.Astrophys.* 37, 103.

—— (1957). *Z.Astrophys.* 43, 161.

—— (1958). *Proc.phys.Soc.Lond.* 72, 263.

RAYCHAUDHURI, A.K. (1974). *J.Math.Phys.* 15, 856.
——— (1975a). *Prog.theor.Phys., Osaka* 53, 1360.
——— (1975b). *Phys.Rev.* D12, 952.
——— and BANERJI, S. (1977). *Phys.Rev.* D16, 281.
——— and DUTTA, A.K. (1974). *J.Math.Phys.* 15, 1277.
REDHEAD, A.C.S. and HEWISH, A. (1976). *Mon.Not.R.astr.Soc.* 176, 571.
REES, M.J. (1968). *Astrophys.J.* 153, L1.
——— (1971). *Mon.Not.R.astr.Soc.* 154, 187.
——— (1972). *Phys.Rev.Lett.* 28, 1669.
——— and REINHARDT, M. (1972). *Astron.Astrophys.* 19, 189.
——— and SCIAMA, D.W. (1968). *Nature,Lond.* 217, 511.
REEVES, H. (1972). *Phys.Rev.* D12, 3363.
———, FOWLER, W.A., and HOYLE, F. (1970). *Nature,Lond.* 226, 727.
———, AUDOUZE, J., FOWLER, W.A., and SCHRAMM, D.N. (1973). *Astrophys.J.* 179, 909.
REFSDAL, S. (1964). *Mon.Not.R.astr.Soc.* 128, 295.
——— (1966). *Mon.Not.R.astr.Soc.* 132, 101.
——— (1970). *Astrophys.J.* 159, 357.
———, STABELL, R., and DE LANGE, F.G. (1967). *Mem.R.astr.Soc.*. 71, 143.
REINHARDT, M. (1972). *Astron.Astrophys.* 19, 104.
——— (1973). *Z.Naturforsch.* 28, 529.
RILEY, J.M. (1973). *Mon.Not.R.astr.Soc.* 161, 118.
RINDLER, W. (1956). *Mon.Not.R.astr.Soc.* 116, 6.
——— (1969). *AStrophys.J.* 157, L147.
RINGENBERG, R. and MCVITTIE, G.C. (1969). *Mon.Not.R.astr.Soc.* 142, 1.
——— ——— (1970). *Mon.Not.R.astr.Soc.* 149, 341.
ROBERTSON, H.P. (1928). *Phil.Mag.* 5, 835.
——— (1929). *Proc.Natn.Acad.Sci.* (*U.S.A.*) 15, 822.
——— (1935). *Astrophys.J.* 82, 284.
——— (1936a). *Astrophys.J.* 83, 187.
——— (1936b). *Astrophys.J.* 83, 257.
ROBINSON, B.R. (1961). *Proc.Natn.Acad.Sci. U.S.A.* 47, 1852.
ROCKSTROH, J. and WEBER, W.R. (1969). *J.geophys.Res.* 74, 5401.
ROEDER, R.C. (1975). *Astrophys.J.* 196, 671.

ROGERSON, JR., J.B. and YORK, D.G. (1973). *Astrophys.J.* 186, L95.

ROLL, P.G. and WILKINSON, D.T. (1966). *Phys.Rev.Lett.* 16, 405.

ROSE, W.K. and TINSLEY, B.M. (1974). *Astrophys.J.* 190, 243.

ROWAN-ROBINSON, M. (1968). *Mon.Not.R.astr.Soc.* 138, 445.

——— (1971). *Nature,Lond.* 229, 388.

——— (1972). *Astrophys.J.* 178, L81.

——— (1974). *Mon.Not.R.astr.Soc.* 168, 45P.

——— (1976). *Nature,Lond.* 264, 603.

ROY, S.R. and SINGH, P.N. (1974). *J.Phys.* A7, 452.

RUBAN, V.A. and FINKELSTEIN, A.M. (1975). *Gen.Relativ.Grav.* 6, 601.

——— ——— (1976). *Astrofizika (USSR)* 12, 371.

RUBIN, V.C., THONNARD, N., FORD, W.K., JR., and ROBERTS, M.S. (1976). *Astr.J.* 81, 719.

RUBTSOV, V.I. and ZATSEPIN, V.I. (1968). *Can.J.Phys.* 46, S518.

RUFFINI, R. and WHEELER, J. (1971). *Esro.Symposium SP52* (eds. A.F. Moore and V. Hardy). Paris.

RUZMAIKINA, T.V. and RUZMAIKIN, A.A. (1969). *Soviet Phys. JETP* 29, 934.

RYAN, M.P. (1972). *Astrophys.J.* 177, L79.

RYLE, M. and NEVILLE, A.C. (1962). *Mon.Not.R.astr.Soc.* 125, 39.

SACHS, R.K. (1961). *Proc.R.Soc.* A264, 309.

——— and EHLERS, J. (1971). *Astrophysics and general relativity*, Vol.2. Gordon and Breach, London.

——— and WOLFE, A.M. (1967). *Astrophys.J.* 147, 73.

SAFKO, J.L., TSMAPARLIS, M., and ELSTON, F. (1977). *Phys. Lett.* A60, 1.

SALAM, A. (1975). *Quantum gravity*. Clarendon Press, Oxford.

SANDAGE, A.R. (1961a). *Astrophys.J.* 133, 355.

——— (1961b). *Astrophys.J.* 134, 916.

——— (1970). *Astrophys.J.* 162, 841.

——— (1972). *Astrophys.J.* 178, 1.

——— (1973). *Astrophys.J.* 183, 743.

——— (1975). *Astrophys.J.* 202, 563.

——— (1976). *Astrophys.J.* 205, 6.

——— and TAMMANN, G.A. (1975a). *Astrophys.J.* 196, 313.

SANDAGE, A.R. and TAMMANN, G.A. (1975b). *Astrophys.J.* 197, 265.

—— —— , and HARDY, E. (1972). *Astrophys.J.* 172, 253.

SARGENT, W.L.W. (1974). *IAU Symposium* 58, p.195. Reidel, Dordrecht.

—— and SEARLE, L. (1966). *Astrophys.J.* 145, 652.

SATO, H., MATSUDA, T., and TAKEDA, H. (1970). *Prog.theor.Phys., Osaka* 43, 1115.

—— —— —— (1971). *Prog.theor.Phys.*, suppl. 49, 11.

SAUNDERS, P.T. (1969). *Mon.Not.R.astr.Soc.* 142, 213.

SAVEDOFF, M.P. and VILA, S. (1962). *Astrophys.J.* 136, 609.

SCHATZMANN, E. (1970). Physics and Astrophysics. CERN Lecture.

SCHIFF, L.I. (1964). *Rev.mod.Phys.* 36, 510.

SCHMIDT, B.G. (1973). *Comm.Math.Phys.* 29, 49.

SCHMIDT, M. (1968). *Astrophys.J.* 151, 393.

SCHRAMM, D.N. (1974). *A.Rev.Astr.Astrophys.* 12,

SCHRODINGER, E. (1939). *Physica* 6, 899.

—— (1957). *Expanding universe.* Cambridge University Press.

SCHUCKING, E.A. (1957). *Naturwissenschaften* 44, 507.

—— and HECKMANN, O. (1958). *La structure et l'evolution de l'universe.* Stoops.

SCHWARTZ, D.A. (1970). *Astrophys.J.* 162, 439.

SCHWINGER, J. (1968). *Phys.Rev.* 173, 1264.

—— (1970). *Particles, sources and fields.* Vol.I. Addison Wesley, Reading, Mass.

—— (1976). *Gen.Relativ.Grav.* 7, 251.

SCIAMA, D.W. (1953). *Mon.Not.R.astr.Soc.* 113, 34.

—— (1962). *Recent developments in general relativity,* P.415. Pergamon Press, Oxford.

—— (1964a). *Rev.mod.Phys.* 36, 463.

—— (1964b). *Rev.mod.Phys.* 36, 1103.

—— (1971). *Relativity and cosmology.* Academic Press, London.

——, WAYLEN, P.C., and GILMAN, R.C. (1969). *Phys.Rev.* 187, 1762.

SCOTT, E.L. (1947)· *Astron.J.* 62, 248

SEARLE, L., and SARGENT, W.L.W. (1972). *AStrophys.J.* 173, 25.

SEELIGER, H. (1896). *Münch.Ber.Math.Phys.Kl.* 373.

SEGAL, I.E. (1972). *Astr.Astrophys.* 18, 143.

SCOTT, E.L. (1957). *Astron.J.* 62, 248.

SEGAL, I.E. (1975). *Proc.natn.Acad.Sci. U.S.A.* 72, 2473.

——— (1976). *Proc.natn.Acad.Sci.* 73, 669.

SEIELSTAD, G.A., SRAMEK, R.A., and WEILER, K.W. (1970). *Phys.Rev.Lett.* 24, 1373.

——— and WEILER, K.W. (1971). *Astr.J.* 76, 211.

SELING, T.V. and HEILES, C. (1969). *Astrophys.J.* 155, L163.

SEN, N.R. (1933). *Proc.R.Soc.* A140, 269.

——— (1935). *Z.Astrophys.* 9, 215.

——— (1936). *Z.Astrophys.* 10, 291.

SHAPIRO, I.I., SMITH, W.B., INGALLS, R.P., and PETTENGILL, G.H. (1971). *Phys.Rev.Lett.* 26, 27.

———, ASH, M.E., INGALLS, R.P., SMITH, W.B., CAMPBELL, D.B., DYCE, R.B., JURGONS, R.F., and PETTENGILL, G.H. (1971). *Phys.Rev.Lett.* 26, 1132.

SHAPIRO, S.L. (1971). *Astr.J.* 76, 291.

SHEPLEY, L.C. (1964). *Proc.natn.Acad.Sci. U.S.A.* 52, 1403.

——— (1969). *Phys.Lett.* A28, 695.

———, DAY, G.A., EKERS, R.D., and COLE, D.J. (1966). *Austr.J.Phys.* 19, 837.

SHIMMINS, A.J. and BOLTON, J.G. (1974). *Aust.J.Phys.Suppl.* 32, 1.

SHIVANANDAN, K., HOUCK, J.R., and HARWIT, M.O. (1968). *Phys. Rev.Lett.* 21, 1460.

SHKLOVSKI, J. (1967). *Astrophys.J.* 150, L1.

SILK, J. (1970). *Space Sci.Rev.* 11, 671.

——— (1973a). *Rev.Astr.Astrophys.* 11, 269.

——— (1973b). Cargese Lectures in Physics, Volume 6 (ed. E.Schatzman) Gordon and Breach, New York.

——— (1974). *Astrophys.J.* 194, 215.

——— and AMES, S. (1972). *Astrophys.J.* 178, 77.

——— and SHAPIRO, S.L. (1971). *Astrophys.J.* 166, 249.

——— and TARTAR, J. (1973). *Astrophys.J.* 183, 387.

SOLHEIM, J.E. (1966). *Mon.Not.R.astr.Soc.* 133, 321.

———, BARNES, III, T.G., and SMITH, H.J. (1976). *Astrophys.J.* 209, 330.

SOLINGER, A.B. and TUCKER, W.H. (1972). *Astrophys.J.* 175, L107.

SOM, M.M. and RAYCHAUDHURI, A.K. (1968). *Proc.R.Soc.* A304, 81.

SPITZER, L., DRAKE, J.F., JENKINS, E.B., MORTON, D.C., ROGERSON, JR., J.B., and YORK, D.G. (1973). *Astrophys.J.* 181, L116.

SRAMEK, R.A. (1971). *Astrophys.J.* 167, 255.

STABELL, R. (1968). *Mon.Not.R.astr.Soc.* 138, 313.

STAUFFER, D. (1972). *Phys.Rev.* D6, 1797.

STECKER, F.W. (1969a). *Phys.Rev.Lett.* 21, 1016.

——— (1969b). *Phys.Rev.* 180, 1264.

——— (1973a). *Nature,Lond. (Phys.Sci.)* 241, 74.

——— (1973b). *Proceedings of International Symposium on X-rays*. (Greenbelt). (unpublished)

——— and PUGET, J.L. (1972). *Astrophys.J.* 178, 57.

———, MORGAN, D.L., and BREDEKAMP, J. (1971). *Phys. Rev.Lett.* 271, 1469.

STEIGMAN, G. (1973). *Cargese lectures in physics*, Vol.6. Gordan and Breach, New York.

——— (1974). *Confrontation of cosmological theories with observational data.* Reidel, Dordrecht.

STEWART, J. (1969). *Mon.Not.R.astr.Soc.* 145, 347.

——— and HAJICEK, P. (1973). *Nature,Lond. (Phys.Sci.)* 244, 96.

STOKES, R.A., PARTRIDGE, R.B., and WILKINSON, D.T. (1967). *Phys.Rev.Lett.* 19, 1199.

STOVER, R., MORAN, T., and TRISCHKA, J. (1967). *Phys.Rev.* 164, 1599.

STRITTMATTER, P.A., BRECHER, K., and BURBIDGE, G.R. (1972). *Astrophys.J.* 174, 91.

SUNYAEV, R.A. and ZEL'DOVICH, Y.B. (1969). *Nature,Lond.* 223, 721.

——— ——— (1970). *Astrophys.Space Sci.* 7, 3.

——— ——— (1972). *Commun.Astrophys.* 4, 173.

SWARUP, G. (1975). *Mon.Not.R.astr.Soc.* 172, 501.

SWINERD, G.G. (1977). *Gen.Relativ.Grav.* 8, 379.

SYNGE, J.L. (1937). *Proc.Lond.math.Soc.* 43, 376.

TAFEL, J. (1973). *Phys.Lett.* A45, 341.

——— (1975). *Acta phys.pol.* B6, 537.

TALBOT, JR., R.J. and ARNETT, W.D. (1973). *Astrophys.J.* 186, 51.

TAUB, A.H. (1951). *Ann.Math.* 53, 472.

TAUBER, G.E. and WEINBERG, J.W. (1961). *Phys.Rev.* 122, 1342.

TELLER, E. (1948). *Phys.Rev.* 73, 801.

THIRRING, H. and LENSE, J. (1918). *Phys.Z.* 19, 156.

THOMSON, I.H. and WHITROW, G.J. (1967). *Mon.Not.R.astr.Soc.* 136, 207.

——— ——— (1968). *Mon.Not.R.astr.Soc.* 139, 499.

THORNE, K.S. (1967). *Astrophys.J.* 148, 51.

THORSTENSEN, J.R. and PARTRIDGE, R.B. (1975). *Astrophys.J.* 200, 527.

TIFFT, W.G. (1972). *Astrophys.J.* 175, 613.

TINSLEY, B.M. (1972a). *Astrophys.J.* 173, L93.

——— (1972b). *Astrophys.J.* 174, L119.

——— (1972c). *Astrophys.J.* 178, 319.

——— (1972d). *Astron.Astrophys.* 20, 383.

——— (1973a). *Astrophys.J.* 184, L41.

——— (1973b). *Astron.Astrophys.* 24, 89.

——— and GUNN, J.E. (1976). *Astrophys.J.* 203, 52.

TOLMAN, R.C. (1931). *Phys.Rev.* 38, 1758.

——— (1934a). *Proc.natn.Acad.Sci. U.S.A.* 20, 169.

——— (1934b). *Relativity, thermodynamics, and cosmology.* Clarendon Press, Oxford.

——— (1949). *Rev.mod.Phys.* 21, 374.

TOMITA, K., NARIAI, H., SATO, H., MATSUDA, T., and TAKEDA, H. (1970). *Prog.theor.Phys., Osaka* 43, 1551.

TRAUTMAN, A. (1972a). *Bull.Acad.pol.Sci.Sér.Sci.math.astr. phys.* 20, 185.

——— (1972b). *Bull.Acad.pol.Sci.Sér.Sci.math.astr. phys.* 20, 503.

——— (1972c). *Bull.Acad.pol.Sci.Sér.Sci.math.astr. phys.* 20, 895.

——— (1973a). *Bull.Acad.pol.Sci.Sér.Sci.math.astr. phys.* 21, 345.

——— (1973b). *Nature, Lond. (Phys.Sci.)* 242, 7.

——— (1975). *Ann.N.Y.Acad.Sci.* 262, 241.

TRECIOKAS, R. and ELLIS, G.F.R. (1971). *Comm.Math.Phys.* 23, 1.

TROMBKA, J.I., METZGER, A.E., ARNOLD, J.R., MATHESON, J.L., REEDY, R.C., and PETERSON, L.E. (1973). *Astrophys.J.* 181, 737.

TRÜMPER, M. (1971). *Astrophys.J.* 164, 223.

TRURAN, J.W. and CAMERON, A.G.W. (1971). *Astrophys.Space Sci.* 14, 179.

TUAN, S.F. (1972). *Phys.Rev.* D6, 1445.

TURNER, E.L. and GOTT, III, J.R. (1975). *Astrophys.J.* 197, L89.

VAIDYA, P.C. (1977). *Pramana* 8, 512.

VAN FLANDEM, T.C. (1974). *Bull.Am.astr.Soc.* 6, 206.

——— (1975). *Mon.Not.R.astr.Soc.* 170, 333.

VAN STOCKUM, W.J. (1937). *Proc.R.Soc.Edin.* 57, 135.

VAUCOULEURS, DE G., (1972). *IAU Symposium 44* (ed. A.B. Evans) Reidel, Dordrecht.

——— (1970). *Science* 167, 1203.

——— (1971). *Publ.astr.Soc.Pacif.* 83, 113.

——— and VAUCOULEURS, DE A. (1973). *Astron. Astrophys.* 29, 1091.

——— (1977). *Nature,Lond.* 266, 125.

VETTE, G.L., GRUBER, D., MATHESON, J.L., and PETERSON, L.E. (1970). *Astrophys.J.* 160, L161.

WAGONER, R. (1967). *Nature,Lond.* 214, 766.

——— (1969). *Nature,Lond.* 224, 481.

——— (1973). *Astrophys.J.* 179, 343.

———, FOWLER, W.A., and HOYLE, F. (1967). *Astrophys.J.* 148, 3.

WALKER, A.G. (1936). *Proc.Lond.math.Soc.* 42, 90.

WAMPLER, E.J., ROBINSON, L.B., BALDWIN, J.A., and BURBIDGE, E.M. (1973). *Nature,Lond.* 243, 336.

WEBSTER, A. (1976a). *Mon.Not.R.astr.Soc.* 175, 61.

——— (1976b). *Mon.Not.R.astr.Soc.* 175, 71.

WEILER, K.W., EKERS, R.D., RAIMOND, E., and WELLINGTON, K.J. (1974). *Astron.Astrophys.* 30, 241.

WEINBERG, S. (1967). *Phys.Rev.Lett.* 19, 1264.

——— (1971). *Astrophys.J.* 168, 175.

——— (1972a). *Phys.Rev.Lett.* 27, 1688.

——— (1972b). *Gravitation and cosmology*. Wiley, New York.

——— (1976). *Astrophys.J.* 208, L1.

VON WEIZSACKER, C.F. (1951). *Astrophys.J.* 114, 165.

WELCH, W.J., KEACHIE, S., THORNTON, D.D., and WRIXON, G. (1967). *Phys.Rev.Lett.* 18, 1068.

WERTZ, J.R. (1971). *Awtrophys.J.* 164, 227.

WEYL, H. (1923). *Phys.Z.* 24, 230.

——— (1930). *Phil.Mag.* 9, 936.

WEYMANN, M. (1966). *Astrophys.J.* 145, 560.

——— (1967). *Astrophys.J.* 147, 887.

WEYSSENHOFF, J. and RAABE, A. (1947). *Acta phys.pol.* 9, 7.

WHEELER, J. (1964a). *Gravitation and relativity*. Benjamin, New York.

——— (1964b). *Relativity, groups and topology*. Gordon and Breach, London.

WHITFORD, A.E. (1971). *Astrophys.J.* 169, 215.

WHITTAKER, J.M. (n966). *Nature,Lond.* 209, 491.

WICKRAMSINGHE, N.C., EDMUNDS, M.C., CHITRE, S.M., NARLIKAR, J.V., and RAMADURAI, S. (1975). *Astrophys.Space Sci.* 35, L9.

WILLIAMSON, JR., K.D., BLAIR, A.G., CATLIN, L.L., HIEBERT, R.D., LOYD, E.G., and ROMERO, H.V. (1973). *Nature,(Phys. Sci.)* 241, 79.

WILLSON, M.A.G. (1970). *Mon.Not.R.astr.Soc.* 151, 1.

WILSON, R.W., PENZIAS, A.A., JEFFERTS, K.B., and SOLOMON, P.M. (1973). *Astrophys.J.* 179, L107.

WOLFE, A.M. (1969). *Astrophys.J.* 156, 803.

——— (1970). *Astrophys.J.* 159, L61.

——— and BURBIDGE, G.R. (1969). *Astrophys.J.* 156, 345.

——— ——— (1970). *Astrophys.J.* 161, 419.

WOODWARD, J.F. and YOURGRAU, W. (1973). *Nature,Lond.* 241, 338.

WOODY, D.P., MATHER, J.C., NISHIOKA, N.S., and RICHARDS, P.L. (1975). *Phys.Rev.Lett.* 34, 1036.

WRIGHT, J.P. (1965). *J.Math.Phys.* 6, 103.

YAHIL, A. and BEAUDET, G. (1976). *Astrophys.J.* 206, 26.

YOURGRAU, W. and WOODWARD, J.F. (1971). *Acta phys.Hung.* 30, 323.

ZATSEPIN, G.T. and KUZMIN, V.A. (1966). *Soviet Phys. JETP Lett.* 4, 78.

ZEL'DOVICH, YA. B. (1962). *Soviet Phys. JETP* 14, 1143.

——— (1963). *Soviet Phys. JETP* 16, 1395.

——— (1964). *Soviet Astr.* 8, 13.

——— (1966). *Soviet Phys.Usp.* 9, 602.

——— (1968). *Soviet Phys.Usp.* 11, 381.

——— (1970). *Comm.Astrophys.Space Phys.* 11, 12.

——— (1972a). *Mon.Not.R.astr.Soc.* 160, 1P.

——— (1972b). *Magic without magic: John Archibald Wheeler, a collection of essays in honour of his 60th birthday.* (ed. J.R.Klauder). Freeman, San Francisco, Calif.

——— (1973). *Soviet Phys. JETP* 37, 33.

——— and NOVIKOV, I.D. (1971). *Stars and relativity.* (ed. Thorne and Arnett) Chicago University Press. Vol.1.

——— and SUNYAEV, R.A. (1970). *Soviet Phys. JETP* 29, 1118.

ZEL'DOVICH, Y.A. B., OKUN, L.B., and PIKEL'NER, S.B. (1966). *Soviet Phys.Usp*. 8, 702.

AUTHOR INDEX

Abbotts, P., 78, 263
Adams, P.J., 212, 252
Ade, P.A.R., 102, 253
Agarwal, P., 28, 257
Aichelburg, P.C., 200, 252
Aldrovandi, R., 23, 230, 252
Alexanian, M., 125, 252
Alfven, H., 227, 247, 252
Alpher, R.A., 34, 100, 135, 252
Al'tshuler, B.L., 242, 252
Aly, J.J., 230, 231, 252
Ames, S., 218, 275.
Anand, K.C., 115, 116, 252
Anile, A.M., 222, 252
Arnett, W.D., 141, 276
Arnold, J.R., 78, 277
Arp, H., 41, 252, 259
Ash, M.E., 165, 275
Audouze, J., 135, 272
Austin, T.B., 67, 252

Bahcall, J., 62, 124, 127, 140 150, 252-3, 259
Bahnsen, A., 138, 256
Baldwin, J.A., 112, 278
Baldwin, J.R., 55, 253
Balkowski, C., 41, 253
Bally, J., 1, 253
Bandyopadhyay, N., 212, 217, 243, 253
Banerjee, A., 95, 253
Banerji, S., 82, 85, 168, 188, 217, 243, 253, 272
Barnes, III, T.G., 23, 275.
Barnothy, J.M., 202, 253
Barrow, J., 142, 253
Barry, G.W., 1, 253
Batakis, N., 90, 253
Batchelor, R.A., 140, 263
Baum, W.A., 23, 253
Beam, W.J., 124, 256
Beaudet, G., 143, 279
Beck, F., 37, 253
Beckman, J.E., 102, 253
Beery, J.G., 102, 253, 254
Behr, C., 85, 92, 152, 253, 259
Bekenstein, J.D., 181, 253
Belinskii, V.A., 92, 151, 154, 156, 253-4
Beltrow-Lopez, V., 243, 263
Bennett, A.S., 235, 254
Berge, G.L., 232, 254
Bertotti, B., 66, 254
Bethe, H., 135, 252
Bishop, N.T., 170, 254
Bjorkholm, P., 105, 260
Black, J.H., 141, 254
Blair, A.G., 102, 254, 279
Blake, G.M., 236, 254
Bleeker, J.A.M., 116, 254
Bludman, S.A., 70, 254
Bogdanova, L.N., 231, 254
Bolton, J.G., 235, 236, 254, 275
Bondi, H., 1, 2, 3, 6, 14, 26, 95, 106, 158, 254
Bonnor, W.B., 97, 207, 212, 254
Bonometto, S.A., 219, 254
Bortolot (Jr), V.J., 104, 254
Bottinelli, L., 41, 59, 253, 254
Bowles, J.H., 105, 254
Boynton, P., 101, 102, 105, 109, 255
Braccesi, A., 235, 255

Bracewell, R.N., 105, 257
Branch, D., 59, 255
Brans, C.H., 113, 164, 166, 241, 255
Branson, N.J.B.A., 236, 263
Brecher, K., 31, 77, 78, 105, 212, 215, 216, 255, 276
Bredekamp, J., 232, 276
Bridle, A.H., 235, 255, 259
Brill, D., 164, 255
Brown, B.M., 63, 265
Brown, G.S., 65, 255
Brown, W.K., 138, 256
Bunner, A.N., 78, 267
Burbidge, E.M., 73, 112, 135, 255, 278
Burbidge, G.R., 70, 73, 78, 105, 111, 135, 255, 276, 279
Burger, J.J., 116, 254
Burke, B.F., 101, 255
Burke, J.A., 41, 259

Caderni, N., 105, 255
Cameron, A.G.W., 141, 224, 255, 277
Campbell, D.B., 165, 275
Campussano, L., 24, 255
Canuto, V., 212, 252
Carleton, N.P., 104, 262
Carlitz, R., 223, 255
Carr, B.J., 71, 112, 224, 255
Carswell, R.F., 112, 256
Cartan, E., 182, 256
Caser, S., 23, 230, 252
Catlin, L.L., 102, 279
Cavaliere, A., 105, 261
Ceccarelli, M., 235, 255
Cesarsky, D.A., 141, 256
Chamaraux, P., 41, 253
Chandrasekhar, S., 94, 256
Chapman, G.A., 167, 256
Chen, C., 162, 256
Chernir, A.D., 34, 69, 218, 221, 256, 258, 269, 270
Chernov, A.S., 88, 265
Chernyshev, V.I., 102, 265
Chibishov, G.V., 110, 218, 256, 270
Chitre, D., 5, 256
Chitre, S.M., 112, 240, 268, 279
Christensen, C.J., 135, 256
Chu, W.T., 124, 256
Churchwell, E., 140, 256
Clauser, J.F., 104, 254, 256
Clemence, G.M., 240, 256
Cline, D., 230, 256
Cocconi, G., 243, 256
Cohen, J.M., 90, 253
Cohen, R.L., 23, 256
Cole, D.J., 235, 257, 275
Coleman, P.L., 78, 267
Colgate, S.A., 142, 256
Collins, C.B., 4, 70, 108, 109, 240, 256
Collin-Souffrin, S., 41, 256
Conklin, E.K., 105, 257
Corey, B.E., 107, 257
Cosmo, de, V., 105, 255
Cowan, C.L., 41, 257
Cowsik, R., 70, 77, 78, 128, 257
Crane, P., 70, 257

Dalgarno, A., 141, 254
Dallaporta, N., 222, 257
Danese, L., 219, 222, 252, 254
Daniel, R.R., 115, 116, 252
Danjo, A., 115, 257
Danziger, I.J., 55, 253
Das, A., 28, 257
Dautcourt, G., 42, 110, 257
Davidson, W., 12, 18, 20, 257, 259
Davies, I.M., 235, 267

Davis, M., 224, 235, 255, 257, 259, 264
Davis, R. (Jr)., 71, 140, 257
Day, G.A., 235, 257, 275
De, U.K., 99, 257
Dearborn, D.S., 162, 257
Debrenberg, A.J.M., 116, 254
Demaret, J., 230, 257
Dennis, B.R., 78, 257
Dicke, R.H., 3, 100, 160, 161, 164, 166, 167, 170, 174, 224, 244, 255, 257-8, 271
Dirac, P.A.M., 158, 258
Dixon, M.E., 224, 258
Dixon, R.S., 235, 259
Dodd, R.J., 240, 258
Doroshkevich, A.G., 4, 90, 121, 217, 258, 261
Drake, J.F., 141, 275
Durney, B.R., 167, 258
Dutta, A.K., 89, 272
Dyce, R.B., 165, 275
Dyer, C.C., 62, 67, 68, 258

Eardley, D., 152, 258
Eddington, A.S., 23, 145, 203, 258
Edeskutz, F., 102, 254
Edmunds, M.C., 112, 279
Eggen, O.J., 224, 258
Ehlers, J., 2, 82, 109, 258, 273.
Eiband, A.M., 105, 254
Einasto, J., 69, 258
Einstein, A., 23, 32, 35, 39, 145, 241, 258-9
Eisenhart, L.P., 8, 259
Ekers, D.O., 165, 268
Ekers, R.D., 235, 257, 275, 278
Ellis, G.F.R., 15, 16, 27, 79, 85, 90, 106, 108, 109, 148, 149, 181, 212, 259, 262, 265, 277
Ellis, R.S., 240, 259
Elston, F., 182, 273
Epstein, R.L., 144, 259
Estabrook, F.B., 85, 259
Euler, H., 196, 259
Evans, A.B., 12, 18, 20, 149, 257, 259
Ewing, M.S., 101, 259

Fabbri, R., 105, 255
Fanaroff, B.L., 239, 259
Fanti, R., 235, 255
Farnsworth, D.L., 149, 259
Felton, J.E., 77, 259
Fennelly, A.J., 65, 259
Field, G.B., 41, 73, 75, 134, 259
Finkelstein, A.M., 176, 273
Fitch, L.T., 235, 259
Fock, V., 213, 259
Fomalont, E.B., 165, 235, 259, 268
Fong, F., 240, 259
Ford, W.K. (Jr)., 107, 273
Fowler, W.A., 135, 142, 255, 259, 272, 278
Frautschi, S., 124, 150, 223, 252, 255, 259, 261
Friedmann, A., 33, 259
Frogel, J.A., 55, 253
Frost, K.J., 78, 257
Frye, G.M. (Jr)., 78, 263, 265
Fujimoto, M., 232, 260
Fulling, S.A., 150, 270

Galton, J.A., 235, 260
Gamow, G., 100, 135, 252
Ganguli, S.N., 230, 260
Gardner, F.F., 232, 235, 254, 260
Gelato, G., 235, 255

Geller, M.J., 23, 250
Geren, P., 109, 258
Geroch, R., 35, 147, 260
Giacconi, R., 105, 261
Gilman, R.C., 242, 260, 274
Ginzburg, V.L., 134, 150, 260
Giovannini, C., 235, 255
Gisler, G.R., 144, 260
Godel, K., 9, 17, 81, 84, 92, 260
Gold, T., 2, 3, 26, 111, 254, 260
Golden, L.M., 235, 260
Goldenberg, H.M., 166, 258
Goldoni, R., 242, 260
Goldwire, H.C., 141, 271
Gorenstein, P., 105, 260
Gott III, J.R., 6, 260, 261, 277
Gougienheim, L., 41, 59, 253, 254
Gould, R.J., 117, 260
Gower, J.F.R., 235, 260
Graaf, de, T., 127, 260
Greenstein, G.S., 172, 260
Greenstein, J., 140, 260
Greisen, K., 114, 116, 260
Grenier, P., 102, 260
Griffiths, R.E., 105, 261
Grischuk, L.P., 90, 261
Groth, E.G., 212, 261
Gruber, D., 78, 278
Grueff, G., 235, 261
Gull, S.F., 105, 261
Gunn, J.E., 45, 55, 56, 62, 67, 70, 71, 73, 260, 261, 271
Gursey, F., 242, 261
Gursky, H., 105, 261
Guth, E., 21, 261

Hagedorn, R., 124, 223, 261
Haggerty, M.J., 6, 261
Hajicek, P., 190, 276
Hainebach, K.L., 37, 261
Hamer, C., 223, 261
Hardy, E., 6, 274
Harmer, D.S., 71, 257
Harnden, (Jr), F., 105, 260
Harries, J.E., 102, 253
Harris, B., 105, 235, 260, 261
Harris, D., 235, 255
Harrison, E.R., 1, 41, 144, 164, 205, 224, 246, 253, 260, 261-2.
Hartwick, F.D.A., 41, 255
Harwit, M.O., 101, 102, 263, 271, 275
Hauser, M.G., 6, 271
Hawking, S., 5, 19, 70, 79, 92, 98, 107, 108, 109, 142, 148, 149, 181, 205, 240, 256, 262
Hawkins, M.R.S., 65, 262
Hayakawa, S., 116, 257
Hayashi, K., 136, 182, 262
Hecht, H.F., 144, 262
Heckmann, O., 11, 15, 17, 18, 19, 20, 58, 85, 86, 88, 90, 151, 262, 274
Hegyi, D.J., 104, 262
Hehl, F.W., 182, 189, 191, 198, 262
Heidmann, J., 24, 41, 255
Heiles, C., 141, 275
Heintzman, H., 162, 262
Heller, M., 28, 150, 262
Henry, P.S., 107, 262
Herman, R.C., 34, 100, 135, 252
Hewish, A., 235, 239, 262, 272.
Heyde, Von der, P., 182, 189, 262
Hiebert, R.D., 102, 254, 279
Hilf, E., 37, 253
Hill, H.A., 167, 262
Hillerbrandt, W., 162, 262

Hinder, R.A., 236, 263
Hively, R., 111, 112, 266
Hoerner, S.V., 239, 263
Hoffman, A., 70, 257
Hoffmann, K.C., 71, 257
Holden, D.J., 235, 263
Hooft, G't., 127, 263
Hopper, V.D., 78, 263
Houck, J.R., 101, 102, 263, 271, 275
Howell, T.F., 101, 263
Hoyle, F., 1, 2, 3, 26, 39, 41, 49, 110, 135, 142, 150, 162, 181, 200, 202, 255, 259, 263, 272, 278
Huang, K., 124, 263
Huchtmeier, C., 140, 256
Huchtmeier, W.K., 140, 263
Hughes, R.G., 236, 263
Hughes, V.W., 243, 244, 263
Huizinga, J.S., 102, 253
Hunter, C., 207, 263

Ingalls, R.P., 162, 165, 175, 275
Ingersoll, A.P., 167, 256, 263
Irvine, W., 212, 263
Isaacson, R.A., 98, 263
Isham, C.J., 199, 263

Jaakkola, T., 41, 59, 264
Jacobs, K.C., 88, 264
Jeans, J., 207, 264
Jefferts, K.B., 104, 141, 264, 271, 279
Jelley, J.V., 117, 264
Jenkins, E.B., 141, 275
Johanson, J.M., 45, 265
Johri, V.B., 217, 264
Jones, B.F., 165, 264
Jones, B.J.T., 205, 222, 264
Jones, B.W., 101, 271
Jordan, P., 158, 264
Joseph, V., 88, 264
Jura, M., 140, 264
Jurgons, R.F., 165, 275

Kaasik, A., 69, 258
Kafka, P., 98, 264
Kantowski, R., 82, 86, 264
Kapahi, V.K., 239, 264
Kardashev, N., 215, 264
Kasner, E., 17, 88, 153, 264
Kasturirangan, K., 78, 264
Katgert, P., 235, 264
Katgert-Merkelijn, J.K., 235, 264
Kaufman, S.E., 60, 264
Kawabata, S.E., 232, 260
Keachie, S., 101, 278
Kellermann, K.I., 235, 264, 270
Kellog, E.M., 105, 261
Kembhavi, A.K., 202, 264
Kenderdine, S., 235, 271
Kennedy, J.E.D., 235, 260
Kerlick, G.D., 186, 189, 191, 198, 262, 264
Khalatnikov, I.M., 92, 151, 205, 253-4, 266
Kibble, T.W.B., 182, 264
Kim, Y.S., 124, 256
King, A.R., 85, 90, 149, 212, 259, 264-5
King, J.G., 1, 265
Kirshner, R.P., 59, 265
Kislyakov, A.G., 102, 265
Klein, O., 227, 265
Klimek, Z., 28, 262
Kobetich, J.E., 77, 78, 257
Kocharov, G.E., 128, 265
Kochel, B., 196, 259
Koehler, J.A., 73, 265
Kohler, M., 86, 265

Kompaniets, A.S., 88, 265
Konstantinov, B.P., 128, 265
Kopczynski, W., 185, 194, 265
Kozlovsky, B.Z., 140, 253
Kraus, J.D., 235, 259, 261
Kraushaar, W.L., 78, 267
Kristian, J., 49, 68, 265
Kruszewski, A., 45, 265
Kuchowicz, B., 140, 182, 183, 188, 191, 265
Kundt, W., 125, 126, 265
Kuo, F., 78, 265
Kuzmin, V.A., 116, 279
Kwak, N., 124, 256
Kwan, J., 59, 265
Lake, G., 105, 265
Landau, L.D., 92, 95, 152, 155, 222, 265
Landsberg, P.T., 34, 63, 265
Lang, K.R., 45, 265
Lange, de, F.G., 54, 60, 272
Lansberg, A., 241, 266
Larson, R.B., 224, 266
Laurent, B.E., 227, 266
Layzer, D., 11, 111, 112, 266
Lebskii, Yu, V., 102, 265
Le Denmot, 59, 264
Lee, H., 124, 266
Lemaitre, A., 31, 266
Lense, J., 241, 276
Lepoole, R.S., 235, 264
Leung, Y.C., 124, 266
Liang, E., 84, 152, 212, 258, 266
Liebes, S., 66, 266
Lifshitz, E.M., 92, 95, 151, 152, 155, 205, 222, 253, 254, 265, 266
Longair, M.S., 41, 110, 236, 239, 259, 263, 266
Lord, S.D., 45, 265
Loyd, E.G., 102, 279
Lucchin, F., 219, 222, 254, 257
Lukacs, B., 28, 266
Lukash, V.N., 4, 90, 258, 266
Luke, S.K., 174, 266
Lynden-Bell, D., 59, 70, 224, 242, 258, 266
Lyttleton, R.A., 1, 254
MacCallum, M.A.H., 5, 85, 108, 121, 259, 266
Mace, B., 78, 263
Machalski, J., 236, 266, 267
Mackay, M.B., 235, 254
Mair, K., 37, 253
Maitra, S.C., 95, 243, 266
Makino, F., 116, 257
Malhotra, P.K., 230, 260
Mal'tshev, V.A., 102, 265
Mansfield, V.N., 162, 266
Marochnik, L.S., 150, 266
Marchant, A., 162, 266
Martin, R., 65, 262
Martiz, T.Z., 102, 253
Maslowski, J., 235, 236, 266, 267
Mather, J.C., 102, 267, 279
Matheson, J.L., 78, 277, 278
Matsuda, T., 218, 219, 267, 274, 277
Matsuo, M., 115, 267
Mattig, W., 60, 64, 267
Matzner, R.A., 149, 175, 217, 267, 271
May, R.M., 62, 253
McCammon, D., 78, 267
McColl, M., 102, 267
McCrea, W.H., 11, 26, 39, 150, 267
McIntosh, C.B.G., 8, 34, 168, 267
McIntyre, P., 230, 256
McKellar, A., 103, 267
McLelland, J., 70, 257

McVittie, G.C., 58, 96, 239, 267, 272
Mehra, J., 22, 267
Meier, D.L., 224, 267
Mejia-Lira, F., 125, 252
Melchiorri, B., 105, 255
Melchiorri, F., 105, 255
Meneguzzi, M., 135, 267
Meszares, P., 224, 267
Metzger, A.E., 78, 277
Meyer, P.G., 116, 268
Mezger, P., 140, 256
Mikuno, E., 115, 267
Miley, G.K., 105, 267
Millea, M.F., 102, 267
Mills, B.Y., 235, 267
Milne, A.E., 2, 11, 267
Milne, D.K., 235, 271
Milton, K.A., 164, 267
Misner, C.W., 4, 5, 17, 39, 92, 121, 122, 150, 151, 152, 156, 267-8
Misra, R.M., 83, 268
Mitton, S., 232, 268
Moffet, A.T., 141, 256
Moran, T., 124, 276
Morgan, D.H., 240, 258
Morgan, D.L., 232, 276
Morganstern, R.E., 162, 170, 173, 175, 268
Morris, D., 232, 260
Morrison, L.V., 162, 268
Morrison, P., 77, 255, 259
Morton, D.C., 141, 275
Motta, S., 222, 252
Muehlner, D., 100, 101, 102, 103, 268
Muhleman, D.O., 165, 268
Muller, D., 116, 268
Murphy, G.L., 28, 150, 268
Murray, S., 105, 261
Nahm, W., 223, 255, 268
Nandy, K., 240, 258
Nanos, G.P., 113, 268
Nariai, H., 150, 175, 178, 218, 268, 277
Narlikar, J.V., 3, 18, 19, 20, 38, 39, 41, 110, 112, 150, 162, 181, 200, 202, 240, 263, 268-9, 279
Natale, V., 105, 255
Nelson, A.H., 233, 269
Nester, J.M., 262
Neugebauer, G., 27, 269
Neumann, C., 17, 269
Neville, A.C., 235, 273
Ng, Y.J., 164, 267
Nickerson, B.G., 70, 269
Nicoll, J.F., 248, 269
Nielsen, A., 138, 256
Nieto, J.L., 24, 255
Nightingale, J.D., 27, 269
Nishimura, J., 115, 267
Nishioka, N.S., 102, 279
Niu, K., 115, 267
Nolt, I.G., 102, 253
Noonan, T.W., 75, 269
Northover, K.J.E., 105, 261
Novello, M., 191, 224, 269
Novikov, I.D., 4, 90, 121, 123, 132, 258, 260, 261, 269, 279

O'Hanlan, J., 167, 269
Oke, J.B., 45, 54, 55, 59, 62, 140, 253, 261, 269
Okun, L.B., 123, 280
Olson, D.W., 212, 269
Omnes, R., 23, 230, 232, 252, 269
Oort, J., 69, 269
Oppenheimer, R.J., 145, 269
Ostriker, J.P., 47, 55, 69, 141, 142, 269

Ozernoiy, L.M., 110, 134, 218, 221, 256, 260, 269, 270
Ozsvath, I., 84, 85, 87, 94, 95, 99, 270

Pacini, F., 111, 260
Pal, Y., 128, 257
Papapetrou, A., 186, 270
Parijskij, Y.N., 105, 109, 270
Park, D., 34, 265
Parker, L., 150, 270
Partridge, R.B., 70, 71, 101, 105, 108, 109, 224, 255, 265, 269, 270, 276, 277
Pasachoff, J.M., 141, 256
Patchett, B., 59, 255
Patrick, T.J., 105, 254
Pauli, W., 145, 182, 270
Pauliny-Toth, I.I.K., 107, 235, 264, 270
Peach, J.V., 67, 252
Peacock, A., 105, 261
Pearson, T.J., 235, 270
Pecker, J.C., 23, 41, 256, 270
Pederson, R.J., 102, 267
Peebles, P.J.E., 6, 23, 69, 100, 130, 131, 132, 135, 136, 212, 218, 223, 224, 260, 261, 269, 270-1
Pelikhov, N.V., 150, 266
Penrose, R., 79, 148, 262
Penzias, A.A., 73, 74, 100, 101, 104, 105, 141, 264, 271, 279
Perko, T.E., 217, 271
Perola, G.C., 105, 267
Persson, S.E., 55, 253
Peterson, B.A., 73, 261
Peterson, L.E., 78, 277, 278
Petrosian, V., 31, 144, 259, 271
Pettengill, G.H., 162, 165, 175, 275
Phillips, S., 240, 259
Pikel'ner, S.B., 123, 280
Pilkington, J.D.H., 235, 271
Pipher, J.L., 101, 102, 271
Pochoda, P., 161, 271
Pooley, G.G., 235, 271
Predmore, C.R., 141, 271
Press, W.H., 71, 223, 271
Price, R.M., 235, 271
Puget, J.L., 23, 218, 219, 230, 231, 232, 252, 254, 269, 271, 276
Puzanov, V.I., 101, 271

Quirk, W.J., 224, 271

Raabe, A., 186, 278
Raghavan, R., 230, 260
Raimond, E., 165, 278
Raine, D.F., 7, 243, 271
Ramadurai, S., 112, 279
Rao, U.R., 112, 264
Rasband, S.N., 109, 271
Raychaudhuri, A.K., 6, 15, 26, 38, 79, 82, 85, 88, 89, 95, 99, 109, 146, 149, 151, 176, 185, 188, 204, 271-2, 275
Reddish, V.C., 240, 258
Redhead, A.C.S., 239, 272
Reedy, R.C., 78, 277
Rees, M.G., 110, 112, 113, 144, 232, 260, 272
Reeves, H., 128, 135, 141, 143, 272
Refsdal, S., 54, 60, 66, 68, 272
Reinhardt, M., 232, 240, 268, 272
Richards, P.L., 102, 267, 279
Riley, J.M., 165, 272
Rindler, W., 38, 63, 272
Ringenberg, R., 239, 272

Roberts, A.P., 23, 270
Roberts, M.S., 107, 273
Robertson, H.P., 8, 26, 272
Robertson, J.G., 235, 267
Robinson, B.J., 73, 265
Robinson, B.R., 88, 272
Robinson, H.G., 243, 263
Robinson, L.B., 112, 278
Robson, E.G., 102, 253
Rockstroh, J., 116, 272
Roeder, R.C., 62, 67, 68, 258, 272
Rogerson (Jr), J.B., 141, 273, 275
Roll, P.G., 100, 101, 273
Romero, H.V., 102, 279
Rosatelli, C., 235, 255
Rose, W.K., 55, 273
Roucher, J., 102, 260
Rowan-Robinson, M., 6, 59, 111, 236, 237, 273
Roy, S.R., 28, 273
Ruban, V.A., 176, 273
Rubbia, C., 230, 256
Rubin, V.C., 107, 273
Rubtsov, V.I., 116, 273
Ruffini, R., 23, 273
Rustad, B.N., 138, 256
Ruzamaikin, A.A., 90, 273
Ruzamaikina, T.V., 90, 273
Ryan, M.P., 175, 223, 267, 273
Ryle, M., 235, 271, 273

Saar, E., 69, 258
Sachs, R.K., 2, 49, 66, 68, 82, 86, 109, 110, 152, 258, 264, 265, 273
Safko, J.L., 182, 273.
Salam, A., 199, 263, 272
Salomonovich, A.E., 101, 271
Salpeter, E., 31, 243, 244, 256, 271
Sandage, A.R., 6, 45, 54, 59, 63, 66, 224, 258, 269, 273-4
Sargent, W.L.W., 41, 140, 141, 274
Sato, H., 218, 219, 221, 267, 274, 277
Saunders, P.T., 88, 274
Savage, P.D., 45, 265
Savedoff, M.P., 207, 274
Schatzmann, E., 23, 144, 271, 274
Schecter, P., 223, 271
Scheepmaker, A., 116, 254
Schiff, L.I., 240, 274
Schmidt, B.G., 147, 274
Schmidt, M., 236, 274
Schraml, J., 105, 271
Schramm, D.N., 37, 135, 162, 257, 260, 261, 272, 274
Schreder, G.P., 117, 260
Schrodinger, E., 213, 274
Schucking, E.A., 11, 15, 17, 19, 20, 60, 85, 86, 88, 90, 151, 262, 264, 270, 274
Schwartz, D.A., 108, 241, 272
Schwarzschild, M., 161, 271
Schwinger, J., 145, 164, 274
Sciama, D.W., 76, 104, 110, 117, 158, 159, 182, 242, 272, 274
Scott, E.H., 73, 271
Scott, E.L., 63, 274
Scott, P.F., 235, 260, 271
Searle, L., 140, 141, 274
Seddon, H., 240, 258
Seeliger, H., 17, 269, 274
Segal, I.E., 248, 269, 274-5
Seielstad, G.A., 165, 232, 254, 275
Seling, T.V., 141, 275
Semeniuk, I., 45, 265
Sen, N.R., 31, 204, 275
Serov, N.V., 102, 265

Shakeshaft, J.R., 101, 107, 263, 270
Shapiro, I.I., 162, 165, 175, 275
Shapiro, I.S., 231, 254
Shapiro, S.L., 69, 144, 275
Sheather, P.H., 105, 254
Shepley, L.C., 92, 149, 152, 217, 259, 267, 271, 275
Shimmins, A.J., 235, 236, 257, 275
Shipley, J.P., 102, 254
Shirafuji, T., 182, 262
Shivanandan, K., 100, 101, 275
Shklovski, J., 31, 275
Silk, J., 31, 77, 78, 144, 212, 215, 216, 218, 221, 255, 275
Singh, P.N., 28, 273
Sinigaglia, G., 235, 255
Sitter, de, W., 32, 35, 258
Smith, H.J., 23, 275
Smith, W.B., 162, 165, 175, 275
Soderholm, L., 227, 266
Sofue, Y., 232, 260
Soifer, B.T., 102, 263
Solheim, J.E., 23, 54, 60, 275
Solinger, A.B., 105, 261, 275
Solomon, P.M., 75, 141, 279
Som, M.M., 99, 275
Spiegel, E.A., 167, 263
Spitzer, L., 141, 275
Sramek, R.A., 165, 275
Srivastava, D.C., 83, 268
Stabell, R., 54, 60, 272, 275
Staelin, D.H., 101, 259
Staib, J.A., 78, 263
Starbunov, Yu. N., 128, 265
Starkovich, K.S., 101, 271
Stauffer, D., 125, 276
Stebbins, R.T., 167, 262
Stecker, F.W., 78, 116, 218, 231, 232, 276
Steigman, G., 124, 232, 259, 276
Stephens, S.A., 115, 116, 252
Stewart, J., 4, 190, 256, 276
Stokes, R.A., 101, 102, 255, 276
Stothers, R., 162, 256
Stover, R., 124, 276
Strathdee, J., 199, 263
Strauss, E.G., 95, 259
Strittmatter, P.A., 78, 112, 256, 276
Strobel, H., 27, 269
Subramanian, A., 230, 260
Sudarshan, E.C.G., 38, 268
Sudhakar, K., 230, 260
Sunyaev, R.A., 105, 110, 134, 266, 276, 279
Suri, A.N., 78, 257
Suszycki, L., 28, 150, 262
Swanenberg, B.N. 116, 254
Swarup, G., 239, 276
Swinerd, G.G., 98, 276
Synge, J.L., 81, 276
Szamosi, G., 174, 266
Szekeres, P., 31, 271

Tafel, J., 190, 191, 276
Taira, T., 115, 267
Tait, W., 23, 270
Takeda, H., 218, 219, 267, 274, 277
Talbot (Jr), R.J., 141, 276
Talureau, B., 102, 260
Tammann, G.A., 6, 45, 59, 273-4
Tanaka, Y., 116, 254, 257
Tananbaum, H., 105, 261
Tandon, S.N., 128, 257
Tartar, J., 77, 275
Taub, A.H., 8, 17, 85, 92, 242, 249, 268, 276

Tauber, G.E., 109, 276
Tayler, R.G., 142, 262
Teller, E., 160, 276
Thaddeus, P., 104, 254, 256
Thirring, H., 241, 276
Thomas, J.A., 78, 263
Thomson, G.B., 78, 263
Thomson, I.H., 6, 95, 276
Thonnard, F., 107, 273
Thorne, K.S., 88, 142, 151, 194, 198, 277
Thornton, D.D., 101, 278
Thorstensen, J.R., 71, 277
Tifft, W.J., 41, 277
Tinsley, B.M., 47, 55, 56, 62, 65, 70, 142, 174, 202, 224, 253, 255, 260, 261, 266, 269, 271, 273, 277
Tolman, R.C., 25, 29, 30, 33, 34, 118, 145, 203, 277
Tomita, K., 218, 277
Toton, E.T., 175, 267
Tovmassion, H.M., 41, 256
Traub, W.A., 104, 262
Trautman, A., 182, 185, 190, 194, 277
Treciokas, R., 27, 277
Tremaine, S.D., 56, 269
Trischka, J., 124, 276
Trombka, J.I., 78, 277
Trumper, M., 2, 277
Truran, J.W., 224, 277
Tsamparlis, M., 182, 273
Tuan, S.F., 124, 277
Tucker, W.H., 105, 275
Tupper, B.O.J., 168, 269
Turner, E.L., 6, 277

Vaidya, P.C., 97, 277
Valladas, G., 230, 252
Van der Kruit, P.C., 105, 267
Van der Laar, H., 105, 235, 264, 267
Van der Meulen, J., 230, 257
Van Flandern, T.C. 162, 175, 277
Van Stockum, W.J., 94, 243, 278
Vaucouleurs, A. de, 41, 278
Vaucouleurs, G. de, 6, 41, 45, 59, 97, 278
Vereshkov, G.M., 150, 266
Vernon, F.L., 102, 267
Vette, G.L., 78, 278
Vickers, D.J., 102, 253
Vigier, J.P., 23, 270
Vigotti, M., 235, 261
Vila, S., 207, 274
Volders, L., 235, 255

Wagoner, R., 111, 135, 136, 137, 138, 139, 235, 278
Wahlquist, H.D., 85, 259
Walker, A.G., 8, 278
Walters, G.K., 141, 271
Wampler, E.J., 75, 112, 278
Wang, C.G., 124, 266
Warren, J.B., 149, 267
Waylen, P.C., 242, 274
Weber, W.R., 116, 272
Webster, A., 236, 278
Weiler, K.W., 165, 232, 275, 278
Weinberg, J.W., 109, 276
Weinberg, S., 27, 62, 68, 106, 113, 124, 127, 128, 172, 263, 278
Weiss, R., 100, 101, 102, 103, 268
von Weizsacker, C.F., 218, 278
Welch, W.J., 101, 278
Wellington, K.J., 166, 278
Werner, M.W., 102, 267
Werner, N.E., 167, 258

Wertheim, G.K., 23, 256
Wertz, J.R., 6, 271, 278
Weyl, H., 9, 278
Weymann, M., 134, 135, 278
Weyssenhoff, J., 186, 278
Wheeler, J., 23, 35, 242, 273, 278
Whiteoak, J.B., 232, 260
Whitford, A.E., 59, 279
Whitrow, G.J., 6, 95, 276
Whittaker, J.M., 39, 279
Wickramsinghe, N.C., 110, 112, 263, 269, 279
Wilkinson, D.T., 100, 101, 105, 107, 108, 224, 255, 257, 270, 273, 276
William, P.J.S., 235, 264
Williamson (Jr), K.D., 102, 254, 279
Wills, D., 235, 260
Willson, M.A.G., 235, 279
Wilson, R.W., 74, 100, 101, 104, 105, 141, 235, 264, 271, 279
Winicour, J., 98, 263
Wolfe, A.M., 70, 109, 110, 111, 273, 279
Wood, C.W., 102, 253
Woodward, J.F., 23, 279
Woody, D.P., 102, 279
Wright, J.P., 94, 256, 279
Wrixon, G., 101, 278
Wu, J.T., 6, 271

Yahil, A., 69, 143, 269, 279
York, D.G., 141, 273, 275
Yourgrau, W., 23, 279

Zatsepin, G.T., 116, 279
Zatsepin, V.I., 115, 273
Zel'dovich, Ya. B., 4, 24, 40, 67, 68, 105, 111, 114, 121, 122, 123, 132, 134, 224, 258, 269, 276, 279-80.
Zieba, S., 236, 266, 267
Zotti, G. de., 222, 252
Zych, A.D., 78, 265

SUBJECT INDEX

α,β,γ theory 100, 135

Absolute velocity and special theory of relativity 106

Acceleration - definition as a 4 vector 80

Affinity- nonsymmetric and torsion, contorsion 182-3

Age of the universe 36, 37, 45

Alignment of spins in Einstein-Cartan theory 188-9, 193

Ambiplasma 227

Angular size 66-8, 239-40

Anisotropy of
  inertia and Mach's principle 243-5
  microwave background
    due to peculiar velocity of earth 106-7
    due to vorticity 107-8
    expressed as temperature anisotropy 106
  universe 4, 39
    and He abundance 142
    and polarisation of background radiation 113
  red shift 42, 107

Antiproton life time 230

Appollo 15 observations 78

Baryon-antibaryon separation due to
  (i) gravitational and magnetic fields 228-9
  (ii) phase transition 231

Bianchi types 4, 5, 85, 87, 88, 90, 108, 249
  type I universe 88, 91, 142
    in Brans-Dicke theory 176-180
    in Einstein-Cartan theory 185-89
    perturbations of density in 217
  type V universe 90, 149
    in Einstein-Cartan theory 191

Bianchi type VII universe 90-1
  VIII universe 92
  IX structure constants 91
    universe, singularity 92, 152-6
    in Einstein-Cartan theory 191

Big bang 9, 28
  models 47

Black holes
  for closure of universe 71
  in galaxy formation 223-4

Bootstrap model 124

Brans-Dicke theory 8
  field equations 163-4
  field equations in revised units 167
  static vacuum solutions 165-6
  universe 168-174
  variational principle 163-4

Bulk viscosity
  and singularity-free models 28
  entropy generation 113
  for massless particles 27
  influence on temporal behaviour 27

C - field and negative energy 150, 181

Cauchy surface 147

Charged Universe 1

Chemical potential
  in Fermi distribution 126
  for electrons, neutrinos and baryons 128-9

Christoffel symbols 21

Class of Riemann space 25

Closed space 25, 35

Closed time like line 17, 94, 99

Closure of space due to
  hidden matter in the corona
    of galaxies 69
  neutral or ionised hydrogen
    71-78

Clumpiness in matter distribution
  effect on deceleration parameter 62
  thermalisation of background radiation 112
  minimum in apparent size 67-68

Condensation in expanding universe background - field
  due to 96

Cosmic electron spectra and microwave background 115

Cosmological principle
  different schools 3-5
  galactic clustering 6
  groups of motion 8
  isotropy of microwave background 4, 6, 109
  Mach's principle 7
  Newtonian cosmology 11

Cosmological term
  in Newtonian theory 17
  in general relativity 23-4
  in Godel universe 92-3
  influence on time behaviour
    of universe 30-32
  quantum fluctuation origin
    24

Curvature scalar 146

Cyanogen molecule as interstellar radiation thermometer 103-4

de Sitter universe 26, 60

Deceleration parameter 37, 42, 45
  effect of evolution 55
  effect of Thomson scattering 62

Density of matter in the universe 5, 6, 69

Density parameter 60

Deuteron abundance 141

Deuteron production theory - observational criterion 142

Dirac's postulate on large numbers 158, 160

Discrete source theory of microwave background 111

Distance in cosmology - different definitions 48-9

Einstein - de Sitter universe 32, 220

Einstein static universe 25, 34, 84,
  instability of 31, 203

Electric type gravitational field 68

Electrically charged or polarised universe 1

Elliptic space 35

Embedding in euclidean space 25

Energy stress tensor for perfect fluids 22

Evolution in galactic luminosities 51-5
  due to variation of G 174

Expansion scalar - definition 80

Extragalactic X-ray sources 105

Faraday rotation and antimatter 232-3

Flat space 35

Foucault experiments and Mach's principle 240

Galaxy formation due to
accretion about black holes 223-4
massive resonance in the hadron era 223
thermal fluctuation of density in
(i) Friedmann universe 207-8
(ii) Lemaitre universe 215
turbulence 222

Geodetic incompleteness 17
as criterion of singular space time 147

Geometry of 3-space 35

Godel universe 84, 92-4
geodesics in 94

Gravitation constant variation
and Mach's principle 158-60
and thermal history of earth 161
limits from observational data 162

Gravitation constant for f-field 199

Gravitational waves 42
effective mass density of 70
influence on dynamics of universe 97-8

Groups of motion for
Godel universe 93
homogeneous universe 85
isotropic universe 7-8

Hawking-Penrose energy condition 148
violation of in case of
(i) different fields 181
(ii) Einstein-Cartan theory 191

Hawking-Penrose theorem 28, 79, 148-9, 180

Hawking-Penrose theorem
- Klein-Alfven theory 230

Helium abundance
at different locales 140-1
influence of temperature fluctuation 144

Homogeneity of matter density and homogeneity of universe 6, 7, 95

Horizons 4, 38, 40
and isotropy of microwave background 118

Hot spot in
microwave background 109
radio sources 239

Hoyle Narlikar theory 200-2

Hubble constant 36, 37, 42, 45, 58, 59

Hydrodynamical approximation 1

Hydrodynamical equation 12

Hydrogen- bounds to cosmic density 73-6

Hyperbolic space 35

Inertial frame
dragging by massive bodies 241
identity with astronomical frame 240
problem in Newtonian cosmology 11, 14

Interstellar grains and thermalisation of background radiation 112

Inverse Compton effect 114

Irregularities in matter distribution limit from microwave background anisotropy 112

Isotropic coordinates:
Schwarzschild solution in 147

Isotropy of universe:
Homogeneity as a consequence 7

Jacobi identity 87

Jeans' condition for gravitational instability 207, 213

Jordan's theory of gravitation 163-4

K correction 53

Kasner universal 7, 88, 153

Kepler's laws of planetary motion 10

Killing vectors 8

Lagrangian for
  Brans-Dicke theory 163
  Einstein Cartan theory 184-5
  electromagnetic field in Einstein Cartan theory 191
  general relativity 164
  Jordan's theory 164
  two tensor theory 199

Large nondimensional numbers 157
  - Dirac's postulate on 158

Large scale in cosmology 2

Leiden frost effect 229

Lemaitre universe 31, 37, 208, 214
  growth of perturbations 214-6
  formation of galaxies 216
  quasi-static phase 214

Lepton degeneracy and Helium abundance 143

Lichnerowicz universe 98-9

Luminosity scatter in galaxies 63

Luminosity volume test for distribution of radiogalaxies 236-7

Mach's principle 7, 9, 35, 108, 158, 200-1, 240-5

Mass variation in Hoyle-Narlikar theory 201

Magnetic fields
  in Einstein Cartan theory 191-
  theory of origin of galactic 224-5

McVittie solution 96-7

Metagalaxy
  contraction and bounce 227
  preferred direction in and isotropy of background radiation 230

Microwave background radiation 3, 4, 6, 19, 39
  polarisation of 113
  step corresponding to 21 cm Hydrogen emission 74
  temperature for different wavelengths 101-2
  temperature in inter-stellar space 103-4
  temperature in the direction of Coma cluster 105

Misra-Srivastava theorem 83, 95

Mix-master universe 4, 5, 116

N-m deceleration parameter,
  from 65
  relation 64
  maximum in 65

Negative pressure 26, 39

Neutrino degeneracy and nucleo-synthesis 143-4
  equilibrium with charged leptons 127
  non-zero rest mass 70-1
  viscosity 4

Newtonian cosmology 10

Newton's law of gravitation 11

Newton's laws of motion 10

Non-uniformity in matter distribution 6, 97, 109-10

Nordstrom's theory 21

Olbers' paradox 3, 245-6

Opacity of universe for photons 117

Open space 35

Optical depth 72

Oscillating universe 33

Particle creation 122, 150

Particle horizon in Universe - see horizon

Perfect cosmological principle 26

Perihelion motion of mercury 10
  and oblateness of sun 167

Perturbation of isotropic universe 203-16
  necessary for galaxy formation 224

Poisson's equation and special theory of relativity 10

Polarised universe 1

Proton energy in cosmic rays- cut off due to microwave background 116

Quantum effects and singularity 150

Quantum fluctuations 24, 224

Quarks in big bang 123

Quasar red shifts 44, 75, 234, 247

Radiation dust universe 33-4

Radiation viscosity
  decay of turbulence due to 222

Radio source distribution
  index in log N-log S plot 235
  possible hole in our neighbourhood 237

Raychaudhuri equation 81, 98, 142, 146, 176, 180, 210
  for Einstein Cartan theory 190

Red shift 23, 36, 40
  anomalous 41
  due to mass variation 167, 201

Reheating of cosmic hydrogen 133-5

Ricci tensor 21
  analogue in Newtonian theory 16

Riemann Christoffel tensor 16, 35, 80

Robertson Walker metric 7-9, 13, 26, 38, 58
  transformation of origin in 118-20

Rotating body in universe 97

Schwarzschild field in universe background 95-7
  line element in isotropic coordinates 147
  singularity 146-7

Scott effect 63

Shear definition 80
  influence on collapse 146, 189
  limit to present value 142
  relation with magnetic field 190, 192

Singular state 33, 36, 45

Singularity
  as geodesic incompleteness 147
  Hawking-Penrose theorem 148
  nature and Hawking-Penrose theorem 148
  of geometric scalars 146
  of infinite density 28
  whimper 149

Sound velocity 218-9

Spherical space 35

Spin - spin interaction
- Einstein Cartan theory 185
enhancement in two tensor theory 200

Spin tensor 184, 186

Static universe 23

Steady state cosmology 3, 26, 39, 66

Stephan's quintet 41

Tachyon 38

Taub - Misner universe 17

Temperature and ionisation of hydrogen 132
upper limit in hadron dominated big bang 124

Tilted universe 212

Uhuru satellite-
discovery of X-ray sources 105

Unified field theory- 145

Universe exhibiting rigid rotation and homogeneity 84
with homogeneity, shear vorticity and expansion 90-1

Variational principle for equations of homogeneous universe 87

Velocity of earth from anisotropy
(i) of red shift 107
(ii) of microwave background 107

Vorticity definition 80
influence on collapse in Newtonian cosmology 17, 18
limit from microwave background observations 108
perturbation - decrease due to expansion 208, 210

Warm big bang 124

Weyl postulate 9, 25

Weyl tensor and electric type gravitational field 68
analogue in Newtonian theory 16

X-ray background 77-8
isotropy setting limits to shear and vorticity 108-9

z 7, 40

Zel'dovich limit 40